Restoring the Land

Restoring the Land

ENVIRONMENTAL VALUES, KNOWLEDGE AND ACTION

Edited by
**LAURIE COSGROVE,
DAVID EVANS AND
DAVID YENCKEN**

School of Environmental Planning
The University of Melbourne

Melbourne University Press
1994

First published 1994

Typeset by Syarikat Seng Teik Sdn. Bhd., Malaysia
in 10.5/12.5 Point Baskerville
Printed in Malaysia by SRM Production Services Sdn. Bhd. for
Melbourne University Press, Carlton, Victoria 3053

Distributed in the U.S.A. and Canada by
International Specialized Book Services, Inc.,
5804 N.E. Hassalo Street, Portland, Oregon 97213-3644
Distributed in the United Kingdom, Europe and Africa by
UCL Press Limited, Gower Street, London WC1E 6BT.
The name University College London (UCL) is a registered trade
mark used by UCL Press Limited with the consent of the owner.

National Library of Australia Cataloguing-in-Publication entry

Restoring the land: environmental values, knowledge and
    action.

Bibliography.
Includes index.
ISBN 0 522 84546 0.

1. Land use, Rural—Australia—Management. 2. Soil
degradation—Australia. 3. Soil conservation—Australia.
I. Cosgrove, Laurie. II. Evans, D. G. (David G.).
III. Yencken, D. G. D. (David George Druce), 1931-   .
IV. Title: Working our land to death?

333.76160994

Contents

Acknowledgements   *vii*
Contributors   *ix*

1 Introduction   *1*

Part I   Values: The Meaning of the Land   7

2 The Futility of Utility   *13*
   WILLIAM J. LINES

3 A New Mind and a New Earth   *22*
   ERIC WILLMOT

4 Justice, Sustainability and Participation as Values
   Implied by the Doctrine of Creation   *27*
   PETER HOLLINGWORTH

5 *Terra Incognita:* Carnal Legacies   *37*
   FREYA MATHEWS

6 Property Rights and the Environment   *47*
   SUSAN DODDS

7 Learning from Other Cultures   *59*
   KWI-GON KIM

Part II   Knowledge: Asking the Right Questions   77

8 Environmental Research Policy in Australia   *83*
   RON JOHNSTON AND TRICIA BERMAN

9 Funding and Conduct of Environmental Research   *93*
   BRIAN FINLAYSON AND TOM McMAHON

10 Science and Environmental Research: A Feminist
Critique   107
JONI SEAGER

11 Politics, Science and the Control of Nature   116
MICHAEL WEBBER

Part III   Action: Dealing with Land Degradation   133

12 Productivity and Sustainability of Agricultural Land   138
ADRIAN EGAN AND DAVID CONNOR

13 Turning Research into Action on the Farm   151
LEIGH WALTERS AND ALBERT ROVIRA

14 The View from the Farm   165
PETER SMALL

15 Alternatives for Achieving Sustainable Land Use   172
JOHN BRADSEN

16 How Government Policy Can Effect Environmental
Change   193
EVAN WALKER

17 Changing Cultures in the Farming Community   202
SHARMAN STONE

18 Conclusion: Environmental Values, Knowledge
and Action   217
DAVID YENCKEN

Notes   237
Bibliography   249
Index   264

Acknowledgements

This book grew out of a conference called *Working our Land to Death? Environmental Knowledge, Values and Action* held at The University of Melbourne in July 1992. The conference brought together academics and other researchers, farmers, foresters, miners, politicians, public servants and environmentalists in an exploration of the way knowledge and values influence land management practice and our approach to land degradation. Not all the papers presented at the conference are reproduced in this book. The book has limited its attention to one aspect of land degradation, that related to farming lands, to sharpen the focus of the case study. It has also added some further material and reassembled and organized the papers so that the basic themes are connected clearly and logically.

The conference was remarkable for the breadth of its debate and the exchange and learning that took place between participants. It showed how a discussion could range from the most theoretical to the most practical, and involve philosophers at one pole and practitioners at the other, and in doing so enrich the understanding of all. It was indeed a justification itself for one of the main themes of this book: the need for and benefit of debate about the nature of the knowledge and values that different people bring to environmental decision-making. It gave the editors great encouragement to assemble the material in a book for a wider readership.

We wish to thank the many people who played a major role in organizing the conference, notably the members of the Steering Committee, Professor W. Charters, Professor A. Coady, Dr M. Connor, Professor A. Egan, Professor M. Webber and Associate Professor D. Wood, and those who worked very hard to bring it into being, particularly Ms Anne Seuling and Ms Judy Young. We are also grateful for the work of Ms Alison Temperley and Ms Michele Burder in typing and assembling material for the book. For his work on the maps and diagrams we thank Rob Rawlings.

Judith Wright's poem, 'Eroded Hills', from her *Collected Poems* (1971, p. 83) is reproduced on page 37 with permission from the publishers, Angus & Robertson.

Contributors

Ms Tricia Berman is a Senior Adviser to the Australian Science and Technology Council (ASTEC) which has produced two major reports on environmental research in Australia.

Mr John Bradsen is Senior Lecturer in Law at The University of Adelaide. He has a particular interest in environmental planning and conservation law, including land conservation and native vegetation issues. He is Chair of the Native Vegetation Council and the South Australian member of the Soil Conservation Council.

Professor David Connor is an agronomist, crop ecologist and simulation modeller in the Faculty of Agriculture and Forestry at The University of Melbourne. He has developed a critical analytical approach to the questions of land management, water use and sustainability of land-use practices.

Ms Laurie Cosgrove is a Lecturer in Environmental Studies in the School of Environmental Planning at The University of Melbourne. Her doctoral thesis, 'Environmental Values', is nearing completion.

Ms Susan Dodds is a Lecturer in Philosophy at the University of Wollongong. She has recently completed her doctoral thesis, 'Persons and Property', and has published work in ethics and bioethics.

Professor Adrian Egan is Professor of Agriculture (Animal Sciences) and Dean of the Faculty of Agriculture and Forestry at The University of Melbourne. From 1988 to 1992 he was also Director of the Centre for Farm Planning and Land Management Research, administered through the Faculty of Agriculture and Forestry.

Dr David Evans is Reader in Environmental Studies and Director of Environmental Studies in the School of Environmental Planning at The University of Melbourne. He lectures and researches in the area of resource management, and is currently supervising several Ph.D. candidates working on sustainable development.

Dr Brian Finlayson is Reader in Geography, Head of Department of Geography, and Manager of the Centre for Environmental Applied Hydrology, The University of Melbourne. He has research interests in fluvial geomorphology and hydrology, land degradation and karst geomorphology, and has a long-standing and close involvement with land management agencies in Victoria such as Melbourne Water, the Rural Water Commission and the Counter Disaster College.

The Most Reverend Peter Hollingworth is Anglican Archbishop of Brisbane and Chair of the Anglican Social Responsibilities Commission. He was formally the Executive Director of the Brotherhood of St Laurence. He remains active in the areas of social justice, poverty and social ethics. This is reflected in his various publications and the committees on which he has served. He was named Australian of the Year in 1992 in recognition of his work in these fields for over thirty years.

Professor Ron Johnston is the Executive Director of the Australian Centre for Innovation and International Competitiveness at The University of Sydney. He has been analysing and contributing to science and technology policy for over twenty years.

Dr Kwi-Gon Kim is Associate Professor of Human Ecology at Seoul University. His research, writing and teaching have concentrated on the relationship between urban planning and environment. He co-ordinated the programme of ecological studies of Seoul urban systems for the Korean Man and the Biosphere Committee of UNESCO, and was recently Chairman of the Symposium on the Comparative Study of Metropolis Ecosystems in Asia.

Mr William J. Lines is the widely travelled author of *Taming the Great South Land,* published in 1991. His next book, *An All Consuming Passion: Origins, Modernity and the Australian Life of Georgiana Molloy,* is to be published in 1994.

Professor Tom McMahon is Professor of Environmental Hydrology and Director of the Centre for Environmental Applied Hydrology at The University of Melbourne, and Deputy Director of the Co-operative Research Centre for Catchment Hydrology. For many years he chaired the Ph.D. Committee of The University of Melbourne, and is currently a member of its Council. His research interests cover surface and groundwater hydrology, irrigation and water engineering, and related environmental issues.

Dr Freya Mathews lectures in philosophy and in women's studies at La Trobe University, and is one of Australia's best-known women philosophers. In recent years, she has turned her attention to feminist issues and to environmental philosophy. Her book, *The Ecological Self,* has just been published.

Dr Albert Rovira is Director of the Co-operative Research Centre on Soils and Land Management in South Australia. He has been widely active in the areas of soil microbiology, cereal root diseases and sustainable agriculture.

Dr Joni Seager is a feminist geographer and environmentalist who teaches at the University of Vermont in the United States. She is the author, most recently, of *The State of the Earth Atlas* (1992) and *Earth Follies: Feminism, Politics and the Environment* (1993).

Mr Peter Small: Land management for sustainability has always been a part of the Small family ethos. Peter follows a family tradition in working the land, and also has a keen interest in research and education. Currently, he is a member of the Regional Advisory Committee for South West Region Department of Food and Agriculture. He is Chairman of the Wool and Rural Industries Skill Training Centre, which has the objective of raising the education and training opportunities of those in the farm sector.

Ms Sharman Stone is a rural sociologist, and was one of the first to study the effects of soil salinization on the community organization and values of Australian farmers. Since then she has managed communications and social planning functions in the Rural Water Commission, the Department of Agriculture and the Victorian Farmers' Federation. She is currently a Research Fellow at Monash University Graduate School of Management.

Professor Evan Walker is the Dean of the Faculty of Architecture and Planning at The University of Melbourne. From 1979 until 1992 he was MLC for Melbourne Province in the Victorian Parliament. From 1982 to 1990 he was a minister in the Cain government. His portfolios included Planning and Environment, Conservation, Agriculture and Rural Affairs, and Arts.

Mr Leigh Walters is a Communications and Technology Transfer Consultant with the Co-operative Research Centre for Soil and Land Management. He has considerable experience in rural extension in the fields of cropping, pasture and farm business management.

Professor Michael Webber has been Professor of Geography at The University of Melbourne since 1985. He is best known as an economic geographer who has written extensively about location theory, regional development and technical change. His work is now mainly concerned with the restructuring of Australian and East Asian industry and the effects of industrial change on immigrants.

Dr Eric Willmot is a distinguished Australian teacher, educational administrator, author and inventor. He was made a Member of the Order of Australia in 1984 for services to education and the field of Aboriginal studies, and he was recently appointed Director-General of Education of the South Australian Education Department.

Professor David Yencken is Professor of Landscape Architecture and Environmental Planning at The University of Melbourne. He was recently appointed President of the Australian Conservation Foundation. He was previously Head of the School of Environmental Planning at The University of Melbourne, Secretary for Planning and Environment, Government of Victoria, and Chairman of the Australian Heritage Commission, Commonwealth of Australia. He has recently carried out research into environmental attitudes of young Australians.

1

Introduction

When Rachel Carson's *Silent Spring* burst upon the world in 1962, it was given an extraordinary reception. Very large efforts were made not only to dismiss its arguments but also to discredit its author. Why should such a concentrated attack have been mounted on the book and the writer? The explanation that is usually offered is that *Silent Spring* seriously challenged powerful vested interests of many kinds, and that those concerned with these interests responded vehemently to defend them. There is, however, a more persuasive explanation: that the response was self-righteous as well as self-interested. *Silent Spring* did indeed pointedly identify organizations responsible for serious environmental pollution. But it also implicitly and sometimes explicitly attacked scientists and the nature of science. It challenged the morality of business and science alike. It was thus an attack on core belief systems. The response was ferocious because it was a defence of core beliefs as well as of interests.

From time to time there are eruptions in environmental debates, and the vehemence of those eruptions is often equally bewildering. We would argue that the passion that is generated in these debates is a reflection of a clash of core values and beliefs. It is the clash between the traditional set of beliefs that have informed and dominated the way Western societies have approached development and progress for many centuries and an emerging set of beliefs based on a new environmental ethic. The significance of these value systems and the influence they have on our approach to the environment is one of the main themes of the book.

Since *Silent Spring* a large body of evidence has been assembled to document the extent and nature of environmental problems. As the evidence has accumulated, so citizen concern has mounted. Remarkable changes have taken place in community attitudes. There are many ways these changes can be depicted. One is through the findings of opinion polls. A poll carried out in Australia in 1979 by Saulwick Age Polls found that only 1 per cent of the population thought 'planning and environment' the most important election issue. By 1989 another poll found that 29 per cent of the population rated 'the environment' the most important issue, compared to 51 per cent 'the economy', 14 per cent 'social welfare', and 4 per cent 'industrial relations'. In 1990 another poll found that 67 per cent felt that Australia should concentrate on 'protecting the environment' compared to only 24 per cent who thought that Australia should concentrate on 'economic growth'. All surveys were carried out by the same pollster.[1] Similar changes in attitudes have been identified in other Western countries.[2]

Together with these changes in attitude, a new environmental awareness amongst consumers and in business, industry and government has been generated. Governments have responded with an array of new legislation and environmental programmes, and with new bodies to administer these laws and initiatives. The physical environmental sciences have blossomed and expanded. From them has come most of the analysis of environmental degradation. From them, too, have come recommendations to deal with the problems that science has identified and analysed, and assumptions about the models needed for change. But all of these responses have failed to achieve the change we need. At best they have been only partially successful. Why?

A large part of the problem has been the underlying assumptions and conceptualization. Conventional wisdom has assumed that as information about a problem is assembled by research and other means it is passed on to those who have responsibility for making decisions about it and those who are directly influenced by it. They in turn modify their behaviour to accord with the new knowledge. But clearly this is not the case. In many situations an environmental problem may be clearly identified and the scientific response to deal with it may be well known, yet little or no effective action takes place.

We are now seeing that there are many things wrong with this conceptualization, even in its most sophisticated forms. One important problem has to do with the concept of knowledge. It is too often assumed that the only knowledge that is valuable is expert 'scientific' knowledge. To a politician this may be the least useful knowledge. What the politician may want is knowledge about the likely effects of a particular action on the community, how the community will perceive the action, how it will judge the government for taking it. To the farmer or business-person directly involved with the problem, many other forms of knowledge may seem much more relevant: for example, knowledge about the impact of the proposed change on the business; the problems of learning new skills and techniques; the response of consumers or the overall market; and the attitudes of financiers. If we are to bring about more effective environmental change, we therefore need a different approach to knowledge. This is the second major theme of the book.

There are other problems with the scientific rationalist approach. It gives little attention to the significance of core values and beliefs and the influence they have on behaviour. It assumes a linear relationship between knowledge, awareness and action. The relationships between values, beliefs, knowledge, attitudes and behaviour are not, however, that simple. Attitudes may directly influence behaviour; but often they do not, because other forces are more influential. Action which is required of us may be the catalyst for change in attitude rather than the reverse. Unless core values and beliefs change, it is unlikely that there will be significant change in behaviour, let alone sustained long term-change. A third theme of the book is, therefore, the interplay of values, knowledge, attitudes and behaviour.

Environmential problems take many forms, including increases in greenhouse gases, ozone layer depletion, land degradation, air and water pollution, and loss of species. The problems have many causes, varying from the local to the global. Water pollution may be caused by a single localized source such as the Chisso factory's pollution of Minamata Bay with mercury-containing effluent.[3] Depletion of the ozone layer and the increase in the carbon dioxide content of the atmosphere are, by contrast, the result of many and varied activities distributed across the globe. Such is the complexity of factors involved that it is difficult to link a specific cause with a direct effect. There are other difficulties: environmental problems can be short-term in their duration, assuming appropriate remedial action is taken, or they can be long-term, persistent and not subject to easy remedy.[4] They are, moreover, embedded in complex economic, social, and political structures which are often very difficult to change. Another emphasis of the book is therefore the complexity of environmental problems.

The aim of the book, therefore, is to explore both concretely and theoretically the linkages between environmental values, knowledge and action. For this enquiry the book concentrates on land degradation and agricultural sustainability, as a case study in the last section of the book, and as a reference point for the discussion of values and knowledge that precedes the case study.

Land degradation is an example of a long-term problem of great complexity. A major national survey into land degradation in Australia carried out by Woods in 1984 for the Australian government found that some of Australia's most important environmental problems were associated with land degradation.[5] Particular problems that Woods identified were: the loss of present and potential production through degradation or irreversible conversion to other uses; land-use conflicts, particularly in coastal areas or near urban centres; water quality and allocations of water supplies; loss of genetic diversity in native flora and fauna; non-sustainable use of living resources, and damage to life-support systems as a consequence of these non-sustainable uses. Although there has been no comparable survey since 1984, it is clear that the situation in 1993 is as bad, if not worse. There is little agreement about how the costs of this degradation should be calculated, but, as Roberts argues, irrespective of the method of calculation used, the costs are very high.[6] The full effects of wide-scale clearing of vegetation or salinization of soils may not even yet be manifesting themselves, making the need for new responses to land degradation imperative.

The Structure of the Book

The first part of the book examines environmental values. The chapters in this part all illustrate the ways in which values permeate our approaches to the land and its use. Each chapter sees the play of values differently. Common to all the views is the assumption that changes are needed to the values that have underlain past Western attitudes to development and 'progress'.

The second part of the book deals with knowledge and the way it is developed through environmental research. Two quite different views are presented. The first assumes that the eventual aim of environmental research is to help us to manage *the* environment more effectively by improving our scientific knowledge base; the second insists that it is *our* environment, that we cannot develop a knowledge base suitable for dealing with environmental problems unless we start with the understanding that we are part of the environment. This second group of authors also argue that environmental problems are essentially political and social, not simply scientific.

The third part of the book examines the specific problem of degradation of farm lands, using the different perspectives of farmers and agriculturalists on the one hand, and of environmental policy makers and researchers on the other. Several programmes developed to counter specific problems of agricultural sustainability are explained to suggest what might be the likely ingredients of successful approaches in the future.

The final chapter draws some conclusions. It first asks whether we can start with the assumption that there is agreement about definitions of environment and sustainability and about the nature and extent of land degradation. Or, is there disagreement about even these definitions and descriptions because of the different values underlying their interpretations?

The chapter then sets out a number of propositions reflecting the main themes of the book, and explores each of these propositions in the light of the case study and other arguments presented in the book. The propositions are as follows:

1. There are different instrumental means of bringing about behavioural change which include: attitudinal change through education and other processes; the use of economic incentives or deterrents; the use of science (knowledge); and

the use of law and administrative process. Some changes may require the influence of one or two of these means only. Others may require all the means working in concert.

2. Those concerned with environmental change tend to perceive solutions as falling within their own expert province, and rarely grasp the importance of other means.

3. Environmental change requires a redefinition of knowledge which embraces local, practical and intuitive as well as expert knowledge.

4. Linear transfers of expert knowledge are not satisfactory means of achieving environmental change. Knowledge exchange needs to be seen as a process of mutual exchange between the users (here the farmers), the researchers and the intermediaries.

5. There is a need for greater status to be given to environmental research as an interdisciplinary activity.

6. Environmental problems are inherently social and political, not simply scientific. Thus environmental research should concentrate as much of its effort on the social and political issues as on the scientific issues, in the same way that environmental decision-making needs to be concerned with social and economic aspects of environmental problems, as much as with scientific aspects.

7. Sustained long-term behavioural change requires a change in core beliefs and values.

8. Dialogue is an essential part of changes in core beliefs and values.

The contributions to the book reveal diverse perspectives on the values that are appropriate for a sustainable future, and on the kinds of knowledge that are most suitable for dealing with complex environmental problems. These perspectives are sometimes contradictory and difficult to reconcile. Further, the positions taken by those advocating changes in values are not easily translated into action of the kind necessary to respond to the demands of a growing world population. The contradictions are themselves an indication of the re-evaluation of core beliefs and values now taking place, a process which has much of its course yet to run.

Part I Values:
The Meaning of the Land

D ebates about environmental problems are contributing to the emergence of a new discourse about values, particularly about the values of Western society; about the contribution of particular values to environmental problems; about the need for change towards so-called 'environmental values' which might be more appropriate for a sustainable future. The necessity for this discourse is well demonstrated in Australia, where European approaches to the land stand in sharp contrast to the approaches of the Aboriginal inhabitants. Australia was settled by a people who saw the land as empty of people and as a resource to be utilized and changed at will. This approach is epitomized in a school social studies textbook written in 1948, which begins its introduction by saying:

> Whether dealing with Australia, the shape of the earth, fire, mechanization or writing we must keep before the pupil the central idea that man has, by using his wonderful powers, gradually made himself dominant on the earth and moulded the environment to his will. This is truly a wonderful thought and the full realisation of this great truth is a part of every child's heritage which must not be denied to him.[1]

The introduction continues with a vignette of the march of civilization across Australia:

> This year you carry on the story of man by following the progress which has been made by him in Australia. You will learn of the difficulties experienced by him in Australia. You will learn of the difficulties experienced in the first settlement, the early quarrels and disappointments, and the slow progress made at first. You will then see how adventurous men set out to explore our forbidding continent and you will learn of the various early settlements. The story is slowly built up—a story of courage and despair, drought and heat and bushfire, and even death. But eventually you will find that man is successful in this strange, new land. You behold a land of promise, a land of waving corn and green pastures and whirring wheels. Man has triumphed at last.[2]

This picture of the development of Australia assumes an alien land, empty of people, but with resources there for the men strong enough to conquer and take them. Missing are, of course, the voices of women, both as generators of heritage and as receivers of even this limited picture of our heritage. Missing too, are the voices of the Aboriginal inhabitants, who viewed the land as imbued with meaning, as sacred.

William Lines begins his paper with the statement that in Australia 'we live on a contested land; we always have'. The contest is a contest of values. William Lines' paper gives an impassioned and uncompromising overview of the conflict between the view of land as a resource and the view of the land as sacred, a conflict still evident in Australian debates over resource use, land rights and heritage. Lines argues that these opposing views are impossible to reconcile, and that attempts to do so lead to duplicitous and dishonest approaches by conservationists. Lines' view is that the only way to avert the problems he foresees in unrestrained growth, and indeed in restrained growth, is to find a new meaning in the land.

The remaining papers in this section search for a new meaning for the land and our relationship with it. Each of the papers, in one way or another, represents a different 'voice' in the discourse about values. The first voice is that dealing with the contribution that so-called traditional values or values held by indigenous peoples may be able to make to action on environmental problems. To some, including David Suzuki,[3] the past harmony referred to by Lines is evidence of a special wisdom from which we in Western society can learn. Eric Willmot critically assesses this argument, suggesting that what he calls First Humanity had a different kind of mind, a mind that we, Second Humanity, no longer possess. It is a futile exercise, then, to seek answers in indigenous cultures and in past practices; rather we must ask questions about our own propensities and find answers in our own capacities. These answers, it is hoped, will lead to a Third Humanity, one more capable of constructing 'a surviving Earth in a new context on an old covenant'. Eric Willmot sees nothing but rhetorical solutions coming from the rhetoric of such things as environmental summits; rather he sees the need for preparedness for a great dying which, if we face it with courage and foresight, will not only save the human spirit but save the planet.

Peter Hollingworth's voice emphasizes the important role to be played by the great religions of the world in developing a cosmology based on values of justice, sustainability and ecological responsibility. He explores the contribution of modern Christian theology to a cosmology able to help us deal with the environmental and spiritual crises of our times through emphasis on the rule of peace and spiritual relationship with nature. Peter Hollingworth urges recognition that our technological capacities are based on a separation from nature, and that our values need to be redirected towards stepping back into nature to care for, protect and nurture what nurtures us. Hollingworth emphasizes the urgency of the need to come together to seek spiritual and ecological meaning.

Freya Mathews is concerned to find ways of living in nature rather than outside it. Her voice is that of the European inhabitants of Australia who are now challenged to assess their superficial relationship with the land. She suggests that it is the failure to 'partake' of the land, both literally and symbolically, which has so inhibited our ability to live within its capacities. European Australians have lived on the surface of the Australian landscape, importing farming methods and bringing with them 'a European Noah's ark'. They have refused to partake of the indigenous food of the land, thereby not partaking of the land in a more fundamental and ultimately more alienating sense. Freya Mathews' paper challenges Australians to think differently about the nature of the farming endeavour, asking us finally to imagine different ways of living, ways that put us back into nature rather than fencing us off.

Western approaches to property and property rights implicitly accept a set of values based on the rights of individuals to use (and abuse) their land in whatever way they wish. Susan Dodds' voice points out the assumptions underlying property rights. She argues that only by spelling out the implications of property rights will we be able to address how property institutions can be restructured to express other values, such as concern for future generations or for nature.

Another voice in the discourse about values is that which sees the values of other cultures, particularly those of Eastern cultures, as able to contribute to more appropriate responses to environmental problems.[4] Kwi-Gon Kim's paper focuses on the contribution that traditional Korean approaches might be able to make to dealing with critical environmental problems. He traces the changes in attitude to developments which have taken place over the last decade in Korea. He points out the richness of understanding and expansion of opportunities for action possible through an alliance between traditional Eastern values and Western knowledge and viewpoints.

It is an illusion to think, as Peter Hollingworth reminds us, that we can be 'value free'. The papers in the section point to the play of values in every aspect of our relationship to the land: in our use of the resources of the land, in our farming practices, in our concept of land as property, and in our sense of divorce from its meaning. All of the papers refer to the role of language in expressing and sometimes in inhibiting the expression of values. Movement into new areas of thinking stretches the capacity of language to communicate. As Hardison expresses it, language forms a 'horizon of invisibility':

> A horizon of invisibility cuts across the geography of modern culture. Those who have passed through it cannot put their experience into familiar words and images because the languages they have inherited are inadequate to the new worlds they inhabit. They therefore express themselves in metaphors, paradoxes, contradictions and abstractions rather than languages that 'mean' in the traditional way—in assertions that are apparently incoherent, or collages using fragments of the old to create enigmatic symbols of the new.[5]

Our language is having to adjust and find ways of expressing the creative, but sometimes painful, efforts to work towards expression of more spiritually satisfying and ecologically sensitive values.

None of the papers in this section clearly delineates what is meant by the term 'values', nor specifically what these new values might be. Together, though, they present a picture of values as complex, many-faceted, and pervasive. The interplay between the messages of the papers clarifies the dimensions of the conflict over values referred to by William Lines, but more importantly shows the way forward. If we are to respond to challenges to incorporate into our value systems a concern for other peoples, other generations, other species and the land itself, then we need to take greater account of the ways in which deeply held values are expressed, and where they are reflected in communication and action. We need personally and collectively to respond to the challenges presented by the papers, to reflect on the meaning we find in the land and the ways in which we might imagine new meanings.

2

The Futility of Utility

WILLIAM J. LINES

We live on a contested land; we always have. Indeed, as evidenced by former President Bush's honest and principled dissent from the fatuous exercise in self-delusion on the part of other world leaders at Rio de Janeiro, we live on a contested planet. In both instances the conflict stems from the same roots, but since the examination of land degradation later in this book mostly refers to Australia, I will confine my comments on the origins and implications of that conflict largely, though not exclusively, to an Australian context.

The conflict in Australia began in 1788. The violence, skirmishes, battles, ambushes, shootings, spearings, rapes, and killings that followed the first settlement were really about land tenure, about the meaning of the earth.

For the Aborigines the land was sacred; it was part of a grand, unchanging, though perpetually renewed, cosmic scheme. The land was their home, their source of life, their Dreaming. They believed that their place in the natural world was central to their being. And, as a consequence of the sacred nature of Aboriginal land tenure, Aboriginal society was technologically satisfied, content, and unprogressive.

For the invaders the land was a prison and, later, a source of profit. British society, even in 1788—the eve of the Industrial Revolution—was technologically restless and driven by an acquisitive impulse, which, formerly condemned as a source of social instability and unhappiness, was going to be seen as a powerful stimulus to economic development.

While members of the First Fleet may have been confused about what they were supposed to be doing in Australia, the men who followed became steadily less so. By the 1820s, leading men in New South Wales were quite clear about the programme of conquest they had to follow. Nature, the earth, was raw material to be turned to the purpose of human convenience. In an 1822 address before the Philosophical Society of Australia, Alexander Berry, a landowner, said:

> It is perhaps happy that Australia's colonisation has been deferred until the present time, when the sum of human knowledge, both moral and physical, is so extended, that attempts to improve the advantages and obviate the disadvantages of nature in this country may be made upon just and rational principles.[1]

And what were these eminently just and rational principles? They revolved around the idea of land as private property and fungible commodity, and around the ideas of utility and efficiency. But these, in turn, are only subsidiaries of a much larger, overarching idea, the idea of progress.

Adam Smith provided the premise for the modern idea of progress. Insatiable desire had previously been regarded as morally suspect; he realized that it could make the world turn and generate material abundance, which, he added, must come before anything else. In Smith's *The Wealth of Nations*, self-improvement assumed the force of moral improvement.

But Smith's insight was incomplete. Avarice needed an ally before the invisible hand and the pursuit of private interest could start generating all that morally sanctioned wealth. The men of the Enlightenment found their ally in science, which, starting with Francis Bacon, had believed the natural world to be devoid of any sacred meaning—that it was, in fact, merely a collection of resources to be consumed. This view denied that humans had any relationship to the natural world beyond use, and denied that humans had any obligations to other organisms. A world devoid of meaning was a world available for human conquest.

To compensate for the loss of meaning in the world, the men of the Enlightenment made a grand promise. The pursuit of profit, together with man's dominion over the planet, would abolish need; and moral improvement, based on the absence of need, would abolish evil and secure freedom. Progress was that simple.

There have been no truer believers in progress than Australians. The idea has virtually dominated Australian history, except for the last twenty or twenty-five years, and has been applied to the Australian landscape with a ferocity and ruthlessness unmatched anywhere in the world.

But there have also been non-believers, even apostates.

First, Aborigines never believed in improvement and an ever-abundant future. They knew the world turned differently from the way conceived by Adam Smith.

Second, many Australians of European descent never accepted progress; for the Western tradition is diverse, and has been and is capable of generating countervailing ideas and beliefs. These dissenters did not need the wisdom of the East or the wisdom of tribal elders to know that something was wrong. An indigenous Western tradition informed our dissenters that insatiable desire was a source of frustration, unhappiness, and spiritual malaise. Our dissenters also understood that the world did not exist solely for human use, that nature was not limitless, and that all life existed under temporal and material constraints.

But the people receptive to these ideas—bushwalkers, bush lovers, a few naturalists and botanists, and some religious folk—never organized. They were, like the Aborigines, the losers in our unfolding progressive story.

Instead of an essentially religious view of life, instead of a view of the land as sacred, Australia acquired, in the early years of this century, conservation and the gospel of efficiency. The conservation movement, far from opposing progress, has always seen itself as an adjunct to, and partner in, progress. From its inception, conservation's mission has been the scientific and efficient use of physical resources, the exploitation of a land devoid of sacred meaning.

As Australia's history has unfolded there have been certain projects, certain built environments, that have served as models and exemplars of a world dedicated to the progressive expansion of human mastery. These undertakings—dams, nuclear explosions, and mines—demonstrate a clear will to dominate the earth.

The cultural implications of the Snowy Mountains Scheme, its value as spectacle and symbol of conquest, its pedagogic value, did not escape the big-thinking men of the time. In 1958, at the Tumut Dam site, Prime Minister Robert Menzies said:

> In a period in which we in Australia are still handicapped by a slight distrust of big ideas and big people or of big enterprise, this scheme is teaching us to think in a big way, to be thankful for big things, to be proud of big enterprises and to be thankful for big men.[2]

Menzies, like many of Australia's leading men, had a fixation with size. Nevertheless, despite his self-infatuation, Menzies understood that the scheme's value was more, much more, than economic (which, in any case, was dubious). The Snowy Mountains Scheme had a display value. This value was directly proportionate to its ability to win prestige for conquest and progress, and build consensus about the proper role for science and technology in the world.

Every school curriculum in Australia included study of the Snowy Mountains Scheme. Throughout the 1950s and the 1960s school children learned that the heroic battle unfolding in the Snowy Mountains was about control—the control of man over nature. Men and machines were reversing the flow of rivers, making them useful and efficient, transforming the nature of Australia and providing for a flourishing future.

Like dams, nuclear explosions in Australia have demonstrated a will to dominate. Howard Beale, then Minister for Supply, described Maralinga as

> a challenge to Australian men to show that the pioneering spirit of their forefathers who developed our country is still the driving force of achievement . . . [Together with Britain] we shall build the defence of the free world, and make historic advances in harnessing the forces of nature.[3]

This pioneering spirit—actually, Adam Smith's 'invisible hand', self-interest and the will to conquer—still govern the destiny of this land and our place on it. Australian governments never tire of demonstrating their commitment to human mastery; without fail they endorse every mega-project proposed

—MFPs, VFTs, enormous pulp and paper mills, massive wood-chip schemes, giant dams and huge mines. Every massive intrusion on the land that promises progress and wealth receives their support.

Most of arid Australia, which is to say most of Australia, is now patterned with seismic lines and other evidence of intrusion, including mines and their associated structures: roads, railway lines, waste dumps, discarded machinery, and the pests and weeds that have followed.

The conflict over Aboriginal land rights, as at the beginning and throughout Australian settlement, revolves around the meaning of the land.[4] The view of the land as a resource, devoid of any meaning beyond profit, conflicts with the view held by Aborigines of the land as a living, sacred homeland, of which they themselves are an inseparable part.

These outlooks are impossible to reconcile. People holding them cannot live on the same land together. The land is either a resource or sacred; it cannot be both. One view must prevail over the other. And progress—the view of a limitless future of consumption—has won out, at least for the moment.

Progress won, in part, because the bush lovers do not employ a sufficiently robust and convincing counter-language. The seismic lines that physically scar this continent reflect, in figurative form, how, over the last two hundred years, Australian society, and more recently, the language of conservation, have been inscribed with the linguistic logic of production and consumption.

Business and industrial metaphors have become so invidious, so pervasive, that it is extremely difficult to talk about the land without the language of development intruding. Management, resources, yield, productivity, utilization, and sustainability represent a debased and biased discourse. This language, imposed from the lexicon of progress, is working our land to death.

Conservationists see no contradiction between progress and conservation. But conservation, both as a political and an environmental strategy, has been a failure, an utter failure—even on its own terms. Despite decades of hard work by well-meaning conservationists, the complex and diverse organism of the earth is in steady decline.

For example, conservation is killing the Great Barrier Reef. This should not surprise. The Great Barrier Reef Marine Park Authority aims not to conserve the reef but to facilitate development—to preserve the reef as a resource. As the authority's first chairman, Graeme Kelleher, said, 'Both conservation and development are necessary, and must be made compatible each with the other'. In other words, the reef must be made, must be remade, to serve the needs of people. And that is a strategy for destruction, not conservation. Less than twenty years after the park's establishment, the reef's present endangered status shows only too clearly the incompatibility of conservation and development. Pollution, loss of biological diversity, and ecological breakdown are the inevitable results of treating the land as a resource.

Conservation fails because its strategies and its language reinforce the primacy of development, of exploitation. Conservation fails because it borrowed its model of human society from Adam Smith and its model of the earth from the scientific programme to conquer nature. The language of conservation, in fact, disavows the world it means to conserve.

Nevertheless, despite a shared vocabulary, a conflict does exist between the advocates of sustainable development and regular developers. Unlike the conflict between miners and Aborigines, however, which is a real conflict because it is about meaning, the conflict between sustainable developers and developers is a false conflict. It is an illusory conflict, because the outcome (a degraded earth) will be the same, whoever wins. The conflict is feigned because for both sides the meaning of the earth—as merely a resource—is the same.

At least the traditional developers understand the nature of the true conflict. George Bush and Australia's John Stone understand, even if conservationists do not, that there is an inherent and irreconcilable conflict between environmental protection and economic growth, between sustainability and development. They understand that if humans do not give way, if economic growth is to continue, then other species will die and the earth will degrade. These interests cannot be conciliated.

Now there are many conservationists in the world, particularly in government, who embrace environmentalism chiefly because they wish to conserve growth, profit, and free markets.

Not all conservationists, however, have been quite as duplicitous. Many, in fact, have sprung from the counter-traditions of the West, which they ignore at their peril. For, without an alternative meaning for the land, genuine conservationists remain linguistically impoverished—trapped by their adherence to the cant of progress.

The most egregious example of the flawed vernacular is the support conservationists lend to that thundering oxymoron, sustainable development. This is a code phrase for a fantasy future in which spiralling consumption leaves no ill consequences.

Conservationists support sustainable development because they are afraid to argue against progress and because they are unable to argue outside the idea of utility. In pursuit of legitimacy, conservationists continue to define the land according to its economic potential. Conservationists continue to employ utility to justify what in reality are moral feelings towards the natural world. And utilitarian rationalizations, like all rationalizations, are readily detectable, sound false, and consequently are unconvincing. I suspect, I know, that many conservationists do not believe their own arguments. They do, however, believe that other people believe utilitarian arguments.

Rainforests, we are told, should be saved because they represent a vast gene pool from which are expected wonderful medicinal and agricultural benefits. But what if they do not? What if most of the natural world proves useless? Are we then free to destroy? Utility leads only to destruction. Utility leads to a world as a purely cultural artefact with no room for non-useful species.

Moreover, utility, from the point of view of nature conservation, is a dead end because no one really believes in it. How can you get passionate about utility? The scientists and conservationists who are actually fighting for rainforests are not at the barricades because they believe rainforests hold miraculous cures. They are there because they love rainforests, because they find rainforests beautiful, because they are in awe of rainforests, because they know rainforests are sacred. Their unwillingness, indeed their embarrassment, to argue explicitly for their passion is the chief cause of their ineffectiveness.

Respect for nature and humility before nature are the only true guarantors of conservation. Otherwise conservation will inevitably fail.

Spurning the idea of progress is but one step. Nature lovers must also reject the welfarist language that handicaps the genuine affection for the natural world that really lies behind environmentalism. Otherwise, the environment remains speechless—the land erodes, salt spreads, forests fall, soil acidity rises, species die, biotas disintegrate, and climates wobble.

There is only one way to halt this destruction; there is only one way to give voice to the environment. Environmentalists must argue with honesty. They must argue something less egotistic, less anthropocentric, less humanistic, and more emphatic. Environmentalists think society is not ready for these arguments. They cynically assume that other people are motivated only by utility and self-interest, and must be appealed to at that level. This is not true. Adam Smith was wrong. The only way to argue for a different, non-exploitative meaning of the earth, the only way to change prevailing and sanctioned modes of thinking and explanation, is to reason outside them. This will give rise to a real conflict.

At the end of the twentieth century, the idea of progress, at least in some intellectual circles, appears tattered, tawdry, and suspect, yet it still claims its adherents. In fact, more people believe in a future of material abundance than ever. Instead of invoking progress directly, today's pathetic defenders of the faith are more circular—they chant jobs. But the premise remains: the boosters still believe in an indefinite expansion of industrial civilization.

Japan, for example, despite its vaunted show of concern at the Earth Summit, plans to double its GNP over the next twenty years. This growth—as it has in the past—can only come at the expense of the planet. To believe otherwise is to indulge in fantasy. To believe simultaneously in growth and sustainability is to be as deluded as the world leaders gathered at the Earth Summit.

To avert the kind of world envisioned by Japan, we must commit to a different future, a future that depends on a different meaning for our land. If the planet is to survive, or rather, if the creatures whose fate we determine are to survive (for the planet will survive no matter what we do and it is only humanistic hubris to think we are stewards of life), it will be through moral courage and heroism.

We need moral courage to recognize and accept the real conflict, and we need heroism—sustained, systematic heroism—to question and overcome the paralysing fatalism that informs the whole discourse of progress. We cannot afford to believe that our future is predetermined by large-scale production and colossal technologies—in short, by development, whether marketed as sustainable or not.[5]

3

A New Mind and
a New Earth

ERIC WILLMOT

The planet Earth is the context of our existence. Take us away from its surface, remove the planet's gravity, and we degenerate physically; remove its atmosphere and we die. Earth is a child of the Universe; we are only a part of that creation. This planet is also the source of our differences. We are a common species, but our nature is determined by the part of Earth to which we belong. Even our appearance is determined by the land of our origin. We have long noses and fair skins if we are born of the snows of Europe. Our skin is black and our hair fuzzy if the tropics were our origin. The slender, durable people of the great Australian deserts were created by them. This places every human in a special relationship with land.

The first human societies on earth seemed possessed of this sense of land. They had a knowing duty of care for the place of their origin beyond modern societies. They believed that land was the context of all of nature. They believed, rightly, that the nature of the trees, the flowers and the beasts, like them, were determined by the lands of their origins.

In many other respects, people of that first world were not unlike us. They did many of the things that we do. They cut down trees, burned fossil fuels and mined the earth. But they did all this within a covenant determined by the nature of the land on which they dwelt. This covenant was such that it ensured a sustainable environmental balance between humanity and the land. This was the perfect balance of the earth's aboriginal worlds, and yet somewhere even in that arrangement there was a tragic flaw.

Some ten to five thousand years ago a new kind of human approach to existence was born. They were people who set aside the ancient covenant. They believed they had found a new context. While First Humanity determined that its social form would reflect the perfection of nature and that it would do the things that nature demanded to ensure its maintenance, Second Humanity believed that social perfection could be found within the human mind and not in nature. Nature was to be simply a beast of burden and they have driven this beast to the point of its death. And we are Second Humanity. We did not come from somewhere else in the cosmos: we are the tragic flaw of First Humanity, for we are their children.

As a general rule, Second Humanity always destroyed the first form wherever it was encountered. But there were some parts of the world where it survived through to the present time. Australia is one of those extraordinary places. Some few of the Aboriginal world survived, protected by the nature of the land they revered. The vast deserts of the interior and the deep complex tropics of the northern point of the continent allowed at least some of them and their social system to survive into the present.

There are other parts of the world where first human societies survived as well. Recently David Suzuki, in his mission for a sustainable future, set out to find from these people the wisdom of the past. He and Peter Knudtsen co-authored a text entitled *Wisdom of the Elders*. As we read this text, we see page by page the sense of the old context and the presence of a knowing duty of care of the land, and the original covenant. Unfortunately the text fails to explain why such wisdom existed then, yet seems not to exist today.

One conclusion is that it is a matter of culture. If only it was so simple. Modern Australian Aboriginal people who have not grown up in the traditional situation are still possessed by a blind duty of care, and this is true of some non-Aboriginal people who grow up on the land. But what of this Wisdom from the past?

In recent consultations with the Pitjatjantjara people of the northwest of South Australia, an education official asked them,

'What is it that you want to do with yourselves?' The answer was simply, 'We want to look after the land'. The question is, does this answer constitute blind culture or a hidden Wisdom?

Suzuki's questions are important to ask but they do not give us an answer, only a statement of the covenant. We can gain something further by exploring the nature of the religions of First Humanity. There we find that land becomes an integral part of human reproduction. We find that land has a spiritual presence and that conception is a process involving this. But this still does not explain why First Humanity was so successful and our second form so much of a failure. First Humanity existed without serious problems for at least ten times the length of our own form. We, on the other hand, have managed to threaten the habitability of this planet.

We need to take a much larger step than Suzuki did to come anywhere near to understanding this fundamental question. In the 1970s in Australia an accidental enquiry of that form was undertaken by Professor Gavin Seagram of the ANU and some of his colleagues.[1] Seagram was a Piagetian psychologist. He selected a group of fully descendant Aranda children who had been brought up in as near as possible to a traditional situation. He examined them for indications of the cognitive steps which Piaget has been able to demonstrate are made by European and other Western children. He found no evidence that these Aboriginal children went through the same process of cognitive development or arrived even at the same point of cognition. Yet he was aware that as adults they were competent human beings.

Perhaps the most important part of his experiment was that he had a control group selected from fully descendant Aranda children who had not grown up in a traditional context. These children were fostered or were placed in institutions, or in some way were brought up by the adults of Second Humanity. He applied the same Piagetian tests to this group and he discovered that there was no significant difference in the outcomes between them and the children of the Western world. These children were the changelings. They had stepped in a single generation from First to Second Humanity.

The vital importance of this work is that it demonstrates quite clearly that culture has a cognitive dimension. Or at least the difference in culture between First and Second Humanity has a distinct cognitive development and difference. It is the thinking base of the two human worlds which is as different as their behaviour, values and attitudes. It is a difference of mind.

So what was it that one human mind was capable of and the other is not? Let us begin our enquiry with a very simple question: we should ask what the people of First Humanity did that was different to what we do today.

As I said earlier, Aboriginal people did cut down trees, they did burn fossil fuels, they did mine the earth; they did not manufacture organochlorides, but neither did early Second Humanity. The significant difference between the two peoples is that First Humanity did not overpopulate the land for which they were responsible. They did not accept that motherhood was always a good thing. More importantly, many of these societies acted as gatekeepers of the future population: they determined which infants should survive and grow into the adult world. In doing these things, they set themselves apart from all other elements of nature, for they did these things knowingly. I do not believe that they were simply controlled by nature. Not only was land the context of humanity, but humanity became the context of the surviving land.

It is not a universally popular view to propose that all of the problems of our environment arise from an apparently uncontrollable and destructive drive to overpopulate. But today perhaps we are desperate enough, all of us, to face the real basis of our faltering destiny. There is, as far as I can see, no special wisdom, no Aboriginal value or attitude, which can reverse the situation which we have fallen into. It was a different kind of mind that enabled First Humanity to do what it did, a mind which Second Humanity does not possess.

If there is to be a way past this dilemma we must ask the questions of Second Humanity rather than our first social form. We must ask what it was that led us to engage in this mindless reproduction which is destroying us. It seems that at least in part it was a quality of the new mind of Second Humanity, arrogance. This was clearly aided and abetted by a sense of immortality, promoted and encouraged by our new religions. In the end, a virtue was made of our ability to colonize and populate the difficult lands of Earth. In modern times we have outgrown much of this. We cannot simply blame religions any more; a sufficient number of us have outgrown at least their simple forms, and many modern societies recognize the problems of populations.

Many societies have taken major steps to escape this destiny. Some nations, like China for instance, are making the attempt, probably too late, and are using quite draconian methods to try to achieve something of the lost covenant. But we are still losing, and losing fast.

The modern rhetoric about new world orders and environment summits is unlikely to discover anything but rhetorical solutions. Nor can the so-called developed or First World any longer help the madly multiplying Third World except through conditions imposed on aid. The people of the Third World can only help themselves. Halve our population and our world has twice the chance of survival. Double it and we all perish.

But if we do nothing, nature will enforce an obvious solution. Nature will bring upon us a great dying. In fact it has already begun. Fairly early in the next millennium it will appear as the most dramatic and horrifying event ever to befall humanity. Unfortunately while a great dying is a solution, it has the potential to destroy the human spirit. The dying is inevitable, but if we face it now and begin to prepare for it our children may begin to develop a different mentality, a different view of our blue Earth. They need to do this, for it is most likely the only world they will ever possess.

First Humanity was destroyed, or else failed when confronted by the new mind of the second human form. The people of Earth need again a new mind, a kind of Third Humanity: something that can construct a surviving Earth in a new context upon an old covenant.[2]

Justice, Sustainability and Participation as Values Implied by the Doctrine of Creation

PETER HOLLINGWORTH

'Values' is a relatively new term covering what the ancients used to describe as 'virtues' or 'goods', in order to signify what was to be regarded as having abiding worth or value. Reference to values is frequently, though not necessarily, tied to religion and religious discourse. Religious values are translated into moral values in a number of different ways. First, they may be regarded as the *source* of moral knowledge; second, as the *sanction* for moral obligation, third, as the means by which moral value choices are *codified* and rationalized; and finally, as a *prophetic critique* in the cause of social justice.[1] A question that is of special interest to contemporary theologians, ethicists and comparative religionists is the way in which these religious and moral values might be applied to the development of ethical guidelines which would help to shape the future of our planet and, more urgently, guarantee its ecological survival.

A New Structure of Meaning

In his early work Paul Tillich addressed the problem of the meaning of religion, and by the time he completed his *Systematic Theology* he talked objectively about God and subjectively about religion in terms of 'ultimate concern'. His Christian existentialist approach effectively linked humanity, existence, religion and God together. Tillich's term for God as 'ultimate concern' brought ethical and religious discourse back to the centre stage of human dialogue, and not before time.

For too long our Western society has made decisions about human beings and the environment with very little reference to such vital considerations. There has been a mistaken view that it is possible to adopt a 'value-free' view of science and technology in relation to the environment and society, which are correspondingly 'value-free'. We now recognize that nothing of significance can be value-free, because every human decision or choice takes place on the basis of some value assumption or other, regardless of any claims to the contrary. (For example, the former view of Australia essentially as a farm and a quarry to be exploited for purely commercial gain has proved to be a grossly simplistic proposition.) Using the concept of religion as ultimate concern, we can construct a paradigm or 'structure of meaning' by which we can view life on our planet as consisting of God as creator, sustainer and renewer; humanity as made in the image of God and called to act as co-partners, stewards and members of the co-operative ecological global community; and the complex interacting web of life which goes to make up the natural order of creation.

Let us begin with God, or, to use the Hebrew, 'Yahweh'. Insofar as the word can be translated, its meaning is 'I am, that I will be'. At the very heart of human and natural existence lies the idea of 'personality' and 'the future', both of which are to be seen as open, dynamic and unfolding concepts. There is nothing whatever in the Judeo-Christian and biblical view of creation which can be regarded as static or fixed. In the Book of Genesis we receive the account of Yahweh, the creator of the universe, the Word, the divine wisdom at work in an evolving creation. Indeed there is nothing contradictory between the doctrines of creation and evolution, because the evolution of the world is not explicable simply in terms of itself. We are told that God created men and women in his own image (the *imago Dei*) to act responsibly within that creation as co-partners and co-workers together in him. When this stage was complete, God looked at it all and saw that it was 'very good', and God celebrated his creation with the Sabbath feast to crown it all.

Now there are embedded here two important values, the one to do with the enjoyment of the created order and the second to do with the stewardship of it. This is summed up in one of our modern Eucharistic prayers which prays, 'You have given us this earth that we might care for it and delight in it and through its bounty you preserve our life'.[2] The subjects of that enjoyment and stewardship are the members of the human race who are the co-partners in creation. There are two key concepts to be noted here. The first, the *imago Dei*, has already been mentioned, while the second is *dominium*, which means that humankind has been given dominion over the earth.

If we go to the Genesis stories about creation and the fall, and to the epic stories like the flood and Noah's ark, there is depicted a situation where people fail to live up to their calling as being made in the image and likeness of God. Thus the divine image is marred, though not obliterated to the extent that the knowledge of good and evil disappears. It is rather the case of knowing the better, but tending to do the worse (Romans 7:18–20). Interestingly, the story of the ark reveals how animals and birds as well as Noah's family are saved, so that they too are part of the ongoing process of creation and salvation, and are therefore to be taken seriously as being important to the created order.

In theological circles there has been a tendency to place too much emphasis upon the salvation of humanity at the expense of the doctrine of creation. Significant efforts are now being made to get things back into a proper alignment.[3] People like Jurgen Moltmann[4] have stressed the importance of developing a sense of community between humans and nature in what he calls 'the community of creation', by developing a 'covenant with nature' in which the rights of humans and of the earth itself are properly balanced. This means that we can no longer view nature as 'unclaimed property'. Such a renewed partnership with nature needs intuitive, imaginative and poetic expressions, as well as scientific and analytical ones.

This takes us to a consideration of the doctrine of *dominium*. Modern critics of the Judeo-Christian tradition point to the charge given in the creation story, 'Be fruitful and multiply and subdue the earth' (Genesis 1:28), as the intellectual foundation of the present ecological crisis caused by the over-populating of the earth and the subjugation of nature to human exploitation. But Moltmann asserts that the biblical concept of subduing the earth has nothing to do with the charge to rule over the world *(dominium terrae)*.[5] The statement is more to do with the dietary commandment that humans and animals are to live from the fruits that the earth brings forth, both plants and trees.

Thus, there is no intention to seize power over nature. The charge 'to rule' is referred to only in Genesis 1:26, 'Let us make humankind in our image, according to our likeness: and let them have dominion over the fish of the sea, and over the birds of the air, and over the cattle, and over all the wild animals of the earth, and over every creeping thing that creeps upon the earth'.[6] But this notion of dominion is related to being made in the image of God, which implies that humanity is to work in harmony with the One who is the creator and preserver of the world. The only 'rule' referred to is that of peace, and, according to Moltmann, human beings are to be like 'Justices of the Peace'. In the second account of the creation (Genesis 2:15) relating to the Garden of Eden, humans (Adam and Eve) are directed to 'till and keep' the garden as the cultivating and protecting work of the gardener. There is no reference here to predatory exploitation. The world is God's creation and it cannot be claimed by men and women. It can only be accepted as a loan and be administered as a trust, according to standards of divine righteousness and justice. As human beings made in the image of God, we have a unique responsibility to participate in developing a cosmology based on justice, sustainability and material well-being, all of which is driven and underpinned by spiritual and communal values on a global scale.

Faith, Science and the Future

To understand the present state of Christian thinking on these matters, it helpful to look at the important work done under the auspices of the World Council of Churches. Especially important was a major international conference held at the Massachusetts Institute of Technology entitled *Faith, Science and the Future*, shorthand for the full title *The Contribution of Faith, Science and Technology in the Struggle for a Just, Participatory and Sustainable Society*.

The significance of that conference, and the two major reports that came out of it, is that it was an attempt by the

World Council of Churches to link its traditional concern for social justice with the emerging concern about ecology. There were three themes that came out of the conference. They are summarized here for use as points of reference. The first was concerned with the relation between science and faith, both of which are forms of human understanding. The role of faith is to determine the right use of science and technology in this very important partnership. The second was to analyse ethical problems resulting from present and prospective developments in particular areas of science and technology. The third was to explore new expressions of Christian social thought and action that are attentive to the promises and the threats of modern science and technology, and at the same time are engaged in a search for a just, participatory and sustainable society.

Australia's Professor Charles Birch was present at the conference. He had come to prominence in these forums somewhat earlier, when he gave a very important address at the General Assembly of the World Council of Churches in Nairobi in the mid-1970s. In that conference he developed eight theses.

1. The world view derived from science really reflects the sort of society in which science is at present practised. He said that the dominant scientific technological world view of today was inherited from a society bent upon mastery over nature, and mirrors its origins. He used the word mastery deliberately, because science has been a predominantly masculine activity.
2. In this view the universe and all that is in it, including living organisms, are conceived as contrivance, which has led to a factory view of nature, with humanity pitted against nature.
3. Christian theology, especially in the West, has accommodated itself uncomfortably to this mechanistic cosmology of science, thus detaching still further nature and humanity and God, and it is now of critical importance that we draw these three together into an interacting whole.
4. This dominant mechanistic world view, whether labelled as theistic or not, has proved itself quite unadapted to our age. It is unecological and it is dehumanizing.
5. This dominant scientific technological world is challenged by personal encounter with the universe in all its wonder and mystery. This challenge is coming largely from outside

science and, indeed, outside Christian theology, though this is changing rapidly because theologians have made some important contributions in the 1980s.

6. We have to accept that challenge, so that we might discover a more relevant ecological view of nature, humanity and God. This requires a change in the relationship between faith and science that has been so dominant in the past.

7. That leads to the seventh thesis, which was that a new partnership of faith and science is beginning to emerge that acknowledges the unity of creation, the oneness of nature, the oneness with nature, humanity and God. It takes seriously both the insights of science and the special characteristics of the human.

8. This ecological view of nature, humanity, and God implies a life ethic which embraces all of life, as well as all of humanity, in an infinite responsibility for all life. Birch says this new ethic and this vision provide a foundation on which to build the ecologically sustainable and socially just global society.

Birch also talked of a new language as a more effective means of sharing and communicating what we must tackle today. Language helps to elaborate and explain the symbols that try to interpret and mediate the values we regard as important in the ordering of society. Such new language must deal with both scientific and theological concepts as a means by which partners in this dialogue can wrestle together with the task of achieving a just, sustainable and participatory society.

It should be remembered that faith and science alike depend upon metaphorical and symbolic language. Both need symbols of purpose and hope, as they set programmes appropriate to future global requirements and constraints. It should also be understood that the spiritual and cultural context of thinking about the cosmos has been through a revolution since biblical times. In the pre-scientific phases of human history, the human race was confronted daily with the forces of nature, which were often threatening and overpowering. Hence, the Genesis injunction to subdue the earth, which must be set in the overall cultural context of the times that tended either to divinize or demonize nature. Genesis teaching began the long historical process which eventually broke the power of blind fatalism by helping people to view themselves as active participants and partners with God in the shaping and subduing of natural forces. Those ancient power relationships have been reversed over time by the development of science and technology. The point has now been reached where, in controlling the forces of nature, humankind has the power to destroy the planet and its species. To respond, we have to look within ourselves as well as at our changing environment. Moltmann began his 1984 Gifford Lectures on the ecological doctrine of creation by saying that the environmental crisis of our times is not only external to humanity, but within human beings themselves. He argued that the crisis could be described as 'apocalyptic', because we are now engaged in 'a life and death struggle for creation on this earth with the forces of nihilism which are preoccupied with the unnatural will to power and an inhumane struggle to dominate the earth, without much concern for the consequences'.[7] The point is that technological invention has now secured the upper hand in dominating nature; it has created a situation the reverse of that of former times, which is of extreme gravity.

This potentially disastrous problem will be rectified only by a set of value commitments which emphasize the relatedness of the creator to the creation, not their separateness. So the dignity of nature as creation is stressed, and humanity's *dominium* is redirected towards and related to the preservation of life on the planet. When human beings intervene in nature, it must be with a care to protect, nurture and restore it. Though we now have the power to transcend nature, we ought never forget that our transcendence is itself rooted and grounded both in God the creator and in nature with its resources, so humanity too remains an integral part of nature, and is never to be separated from it.

A Reinterpretation of Christian Traditions

Let us return again to those two value concepts of the *imago Dei*, that of human beings made in the image of God, and the *dominium*, meaning human dominion over the earth. The notion of *dominium* is always tied to the doctrinal value of the stewardship of the earth. It also allows us to view humanity as both maker and cultivator. Men and women are makers, in that they may with care use nature as raw material and as an instrument. This conception accords with the role of humanity in traditional science and traditional technology. But men and women are also cultivators, in that they preserve and develop the life of the earth. In this view, instead of confronting nature, humans share as partners in its life and accept responsibility for the sustainability of the total system. This conception has an affinity with certain biological conceptions that view evolution primarily as a process which has brought forth humanity. We need a view that integrates these two approaches, a view which sets a limit to the role of the maker and opens the way for the creative imagination, a view in which the reshaping of nature is embedded in co-operation with nature. Such creative and co-operative relationships with nature can be seen as a parable of the work of the creator, and that too is included in the idea of humanity being made in the *imago Dei*, the image of God.

The need to protect and preserve nature should not allow us to forget that, according to the biblical account, nature may be reshaped by humanity and adapted to human needs. In this respect the situation of the rich countries of the earth is completely different from that of the poor countries. In the rich societies, with their preoccupation with growth, nature has been so thoroughly reshaped that many people now live in an artificial world which is alienated from nature. Poorer, traditional societies, which have not yet sufficiently adapted nature to meeting human needs, have been preoccupied with maintaining their equilibrium. In these societies there needs to be a significant development in appropriate technology, but not necessarily in the pattern of the rich countries. The point is that, if we accept the fact that we are created and are therefore creatures, we are, above all, receivers. In modern times, and to the detriment of all creation, this dimension of our relation to God and nature has been hidden by the one-sided emphasis of 'humanity at work'. To counter this, it must be emphasized again that in making and cultivating we are receivers, required to exercise stewardship. This view helps to relativize the importance that we attach to achieving things as the means of fulfilment in life. If such a change in values were to be taken seriously, it would enable the production process to be organized in a more humane way, and the exploitation of nature would also be diminished. The image of God and the dominion of the earth are in fact the birthright of all human beings. But underpinning them both is the commitment to stewardship and accountability. It is this element that we have lost or at least have neglected, and must now rapidly recover. It is a matter of the greatest urgency that it be built into our ethical codes and public policies. Modern theologians like Jurgen Moltmann have sought to elaborate these basic biblical themes, relating them to the environmental crisis of our times. From a Christian perspective this will greatly help in putting the matter in sound perspective, helping humanity to recover its bearings and its balance.

Returning to the values of justice, participation and ecological responsibility, these represent an indissoluble unity. They embrace those who suffer today and those who are yet to be born, and they include as well the non-human aspects of creation. Although our priority today is the defence of those who suffer now, this does not mean that we can reject our responsibility for the earth which is our common home, and which we are to hand on to our children and to their children. The life both of humanity and nature is jeopardized by the same basic attitude of oppressive exploitation which has characterized the past two hundred years of the Industrial Revolution. In the long run, a just society will be a sustainable society. Efforts to preserve ecosystems for the rich and powerful at the expense of the poor and weak may cause the oppressed to resort to revolutionary means to bring about change. This, as we have seen, exacerbates the ecological situation.

Justice and sustainability are more likely to be achieved by the full participation of people at all levels. Hence the importance of emphasizing in the political process the need for a just, sustainable, participatory society. If restrictions on consumption and changes to lifestyle are imposed from above, the response is likely to be one of evasion, a nihilistic attitude towards the future, and the erosion of the human responsibility. The only satisfactory answer is to develop more participatory structures, capacities and motivations in people.

Bodies like the great world religions, including the Christian churches, together have a very important role in working as partners with all other people of goodwill who seek to preserve the earth and all that lies within it. There are fundamental value questions at stake which will be ignored at our peril and at the peril of our world.

5

Terra Incognita: **Carnal Legacies**

FREYA MATHEWS

These hills my father's father stripped,
and beggars to the winter wind
they crouch like shoulders naked and whipped—
humble, abandoned, out of mind.

Of their scant creeks I drank once
and ate sour cherries from old trees
found in their gullies fruiting by chance.
Neither fruit nor water gave my mind ease.

I dream of hills bandaged in snow,
their eyelids clenched to keep out fear.
When the last leaf and bird go
let my thoughts stand like trees here.[1]

My purpose in this chapter is to say a little about the attitudes and values that have underpinned the way we Europeans have related to this land in the last two hundred years, since European settlement began. These attitudes and values explain why Judith Wright's grandfather stripped and wounded the hills of her birthplace. I shall also suggest an alternative philosophical outlook which might lead to a more sensitive relation to the land, and perhaps to a more sensitive and rewarding way of life —an outlook which would restore the leaves and birds to the hills, and would provide food that would satisfy and fulfil the poet's need.

The attitude to the environment that currently prevails in the industrialized world is aggressively *anthropocentric* (i.e.

human-centred). It takes humanity as the only proper object of
moral concern, and it justifies this exclusivity by appeal to the
assumption that human beings stand in some sense apart from,
and above, the rest of Nature. The roots of this anthropocentricism lie in the philosophy of classical Greece—and may
perhaps be traced back even to the Neolithic, as I shall explain
later. In Greek philosophy human identity was divided into a
'higher' mental component and a 'lower' bodily component.[2]
Humanity was distinguished from the rest of Nature by its possession of reason or mind, and was eventually seen as categorically distinct from and superior to the natural world on account
of this. The natural world existed, from this point of view, merely
as backdrop to the drama of human life, in which all metaphysical and moral significance was invested. This assumption
of humanity as the locus of all meaning and value was reinforced
by Christianity,[3] and finally clinched by the new science of the
seventeenth century.[4]

From the viewpoint of the new science, the natural world
was a world both atomistic and mechanistic. That is, it was a
world of mere matter, of aggregates of atoms—discrete par-
ticles, units of substance, logically independent of one another,
held together not by any mutual affinity or intrinsic dynamism,
but by blind, external laws of motion. Since the natural world
was drained of all attributes associated with mind in this way,
mind was clearly no part of such a system. In this world of distinct existences, in which nothing participated in the identity
of anything else, mind constituted a distinct category. Since it
was mind alone which was the repository of telos and agency,
it was mind which was the exclusive locus of value; matter as
the inert and dead, the purposeless and blind, possessed only
the value or meaning that we projected on to it. It followed that
this 'disenchanted', dead world of matter was a fit object for
human use, since it was beyond the reach of moral concern; it
could properly be seen as nothing but a reservoir of resources
for humankind.[5]

The scientific outlook which gave rise to this attitude to
the natural world also facilitated and encouraged the development of the technologies needed to bring it into effect. The
nineteenth and twentieth centuries consequently witnessed

global exploitation of the natural world on a hitherto un-imagined scale.

Many people today are beginning to question anthropo-centrism and the dualistic world view on which it rests—the world view that divides reality into mind (the province of the human) on the one hand, and inert matter (the province of Nature) on the other. This world view has come into question not merely on account of the ominous ecological (and some would add, social) crises to which it has given rise, but also on account of new developments within science itself.[6] A whole spate of new ecophilosophies has emerged, all affirming the intrinsic interconnectedness of the world, all portraying the world as a web or field of shared, interpenetrating essences. By insisting that everything participates in the identity of everything else, these philosophies are reinvesting the world at large with qualities such as life, telos, spirit, and agency. Such a 're-enchantment' of Nature aims to heal the split between mind and matter, animating the world with a quality analogous to mind, emphasizing the intelligence immanent even within our own corporeality. In this way meaning is restored to the world; indeed the world itself, far from being the mere backdrop to the human drama, becomes the primary locus of meaning, from which the meaning of our own lives and those of all other beings is derived. Moreover, the healing of the split between mind and matter, humanity and Nature, re-establishes our affinity, our kinship, with the natural world.

Two examples of this new trend in ecophilosophy are Deep Ecology and ecofeminism.[7] Both of these philosophies reject the dualistic world view that underpins anthropocentricism, and affirm instead a view of the world as intrinsically internally inter-connected and imbued with a meaning-giving principle of its own. Deep Ecology asserts the intrinsic value of all non-human beings and their entitlement to our moral consideration, and also reminds us that our own path to self-realization lies in our ability to recognize our unity with all of life. Ecofeminists point to connections between the domination of nature by humans and other forms of political and social domination. They em-phasize our underlying kinship with the rest of life, and enjoin us to embrace our wider family, promising as our reward an end to estrangement and alienation, a recovery of our sense of being at home in the world. Both philosophies, however, caution that an intellectual appreciation of ideas such as metaphysical inter-connectedness, intrinsic value, biocentric egalitarianism, and so on, will not suffice to transform our relationship with the natural world. We also need to *care*. And care is, at least initially, rooted in the immediate and particular, the local and familiar: we cannot come to care, all at once and in accordance with a change of world view, for life in general. We have to start with what we know. According to the Deep Ecologists, this means identifying with our *place*, digging in and truly dwelling in it, understanding its ecology and all the other aspects of its character. For eco-feminists too, care is cultivated by closely and patiently attending to and empathizing with the non-human beings in our own biotic neighbourhood.

It is at this level—the level of our own place or ecological community—that I think we Europeans here in Australia are doubly alienated from Nature. For while anthropocentricism is currently, as I have remarked, the grand presupposition of in-dustrialized cultures worldwide, and in that sense not peculiar to us, it would be hard to find a people anywhere on earth more alienated from its own land than we are. We have, as a people, no deep sense of being at home here, and no ancestral identification with, or pride in, the original land. It is perhaps this additional dimension of alienation that explains why en-vironmental destruction has in the last two hundred years been more rapid and severe here than almost anywhere else in the world. It is to this problem of our own place that I now want to turn.

Why Australia appears so alien to European settlers is, I think perfectly obvious: this is predominantly an arid land with a highly distinctive topography and biotic community. *That* we are alienated from it is evidenced in a thousand ways, most obviously perhaps in our clinging to the edges of the continent, avoiding its 'dead heart'.[8] However, this alienation was exemp-lified for me in a particularly striking way recently by one of my own personal experiences. Two years ago, when I was camping in outback Western Australia, I encountered my first quandong tree. It was laden with its red fruit, and what a beautiful sight it was in that desert landscape! I picked and ate some of the fruit, and roasted the kernels back at the campfire. And I realized with a shock that this was the first time I had ever eaten of the wild, native food of this land. I had reached mid-life in this country without ever coming to know the taste of Australia.

My experience, which I do not think is atypical, testifies to a categorical cultural refusal genuinely to partake of the land or to 'know' it in a carnal sense. Our colonial forebears brought with them their entire biological—and hence dietary—repertoire of plants and animals. They came here, figuratively speaking, in a European Noah's ark, and we have for all practical purposes remained within that supply ship.[9] It is this denial of the land as a source of food that I want to explore in the remainder of this paper, because while it is by no means the only way in which we have perpetuated our alienation, it is a potent and, I think, unconscious one: by refusing to be nourished by the land we are foreclosing any possibility of regarding it as a *mother* land, and hence as a homeland.

This refusal biologically to partake of the land is a refusal to connect with its essential nature, its spirit; if you like, its *genius loci*. For the essence of the land is incarnate in the life which grows from it—in its native inhabitants. The native fauna and flora absorb the elements of their place—its light, water, rocks, minerals and so on—and evolve into forms that embody the exact balance or pattern of elements present in that place. Because these life forms and all the other elements of the place manifest aspects of the same essence, they are all of a piece: everything that belongs to the place has an affinity with, or partakes of the quality of, everything else. This shared essence is particularly striking, I think, in Australian landscapes: lizards, rocks, gumtrees, emus, kangaroos, all rise like permutations of the same theme, emanations of the land itself.

By refusing to partake, in a visceral way, of the indigenous food of the land, we symbolically signal our refusal to assimilate and be assimilated by its spirit. This symbolic stand may even have literal implications if the notion of spirit is interpreted less metaphorically than I have been interpreting it so far. For while 'spirit' may be understood in this context to denote, in a metaphorical way, the distinctive material character, or essence, of a place, a spirit of place is often attributed to Australia by people who have not had a chance to compare it with other parts of the world, and who are not therefore in a position to appreciate its distinctiveness. Aboriginal people, for instance— past and present—testify to experiencing a powerful sense of the spirit of this land. Their traditional cultures were pre- occupied with spiritual topography. This might be taken to suggest that the spirit of the land is more than a metaphor, that the corporeality of rocks, sand, sky and so on, does indeed pos- sess, in a literal sense, a spiritual dimension.

The idea of spirit immanent in matter, of course, underlies the oldest religious impulses of humankind. It is currently making a comeback on some of the contemporary frontiers of thought, in physics, systems theory and theology, for instance. This idea is often, as I explained earlier, taken to inform, in one way or another, the new world view on which eco- philosophies such as Deep Ecology and ecofeminism are pre- mised. But if we accept some such version of the old doctrine of the *anima mundi*, then it presumably follows that spirit is present in all places. We might wonder why certain places, and indeed certain lands, are picked out as especially exemplifying this universal spirit. Is it the case that certain constellations of elements produce greater spiritual power or intensity than others? It may not be entirely coincidental that humans of out- standing creative intelligence are described as geniuses, while places with certain special features are ascribed with a *genius* also. Just as creative intelligence is present in all humans, yet manifested to a higher degree in some, so *anima mundi* may be present in all matter, yet manifested with greater intensity in certain configurations.

If this is accepted, then it may well be that the distinctive pattern of elements in this continent has resulted in a particu- larly powerful spiritual presence here. Indeed, when we consider the profile of the land—its great antiquity, its spareness and austerity, its power to create an amazing richness and subtlety of life out of material poverty—it is striking how it exemplifies all the qualities we traditionally associate with spirituality!

If it is true then that the spirit of the land is manifested in its native life forms, then our rejection of indigenous nourish- ment and our reliance on European and other non-indigenous foods may not only signal in a symbolic way our refusal to as- similate and be assimilated by the land: it may have pre-empted mutual assimilation at an even deeper, scarcely understood but more powerful, level.

Our refusal to eat indigenous foods may have kept us strangers in a strange land not only by alienating us from the spirit of the place in the way I have described, but also by blinding us to the reality of the land itself, preventing us from perceptually entering into it. To understand what I mean by this, we need first to consider some aspects of the nature of perception. Our powers of perception are in the service of our vital interests. This is their purpose; they do not exist for aesthetic or contemplative purposes, but to alert us to signals relevant to our survival, signals such as those pertaining to danger and to the presence of food. (Hence we are capable of perceiving matter, for instance, but not magnetic fields.) Moreover, within a given perceptual field, our senses discriminate and highlight those features which are salient to us—those which incorporate information of relevance for our appetites and other instincts. In this sense it could be said that it is through our appetites and instincts that we see, hear, smell, and so on. If we inhabit a landscape which we regard as barren of food, then, we will not perceive it in the same searching and discriminating way as those, such as the traditional Aborigines, who regard it as replete with provender. The eye of the Aboriginal perceiver will be drawn into the detail and depth of the landscape; it will probe behind the external appearances into a honeycomb of interiors astir with secret activities. The eye of such a perceiver, guided by appetite, will leap from focus to focus, a thousand times over, and will uncover multiple layers or dimensions. To the perceiver whose appetites are satisfied in other ways, in contrast, the landscape will lack any particular point of focus, and will present a bland, merely two-dimensional surface. It will function, at best, as 'scenery'.

I am reminded in this connection of Advent calendars, whose Christmas scenes from Northern Europe are studded with little doors that open into secret interiors. The indigene in search of food also sees the land, figuratively speaking, as filled with inviting and accessible little doors. For the European, the doors are not only closed; they are themselves invisible. The European is in this sense locked out of the land and, worse still, is not even aware of being so.

So, appetite opens the eyes, and thus reveals the land to the perceiver. We Europeans, repairing to our figurative ark at mealtimes, are walking blind through a closed landscape, capable at most of admiring the 'scenery'. This blindness not only diminishes us, and pre-empts any true understanding of the land; it also reinforces the destructiveness that already flows from our alienation. For the failure to appreciate the full dimensionality of the landscape naturally leads to an actual reduction of that dimensionality. By this I mean that when we fail to recognize the qualitative dimensionality of living spaces, the endless iterations of interiority that life creates, then we feel no compunction about treating those spaces as if they were indeed qualitatively dimensionless—and in reality rendering them so.

This fact was bitterly impressed on me when I visited the site of my childhood home some years ago. I grew up on a three-acre property on the rural fringe of Melbourne, but the area was zoned 'heavy industrial' after my family left. My memory of our home is of a multi-dimensional space, consisting of large gardens, an orchard, vegetable patches, poultry runs, paddocks with old gum trees, an overgrown creek where our ducks and geese sought sanctuary from us, and so on. When I returned, the entire site was covered with a concrete factory, and the creek at the back had been converted into a sealed, straight-edged drain. My over-riding impression was of a massive loss of dimensionality: the space had been dramatically reduced, and had, by virtue of its straightnesses and flatnesses, been made to reflect and hence to confirm our Euclidean model of extension.

My argument so far has been that our failure to understand and feel a part of this land, with all the destructive consequences that flow from this, is both expressed in and perpetuated by our refusal to acknowledge and partake of native foods.[10] This alimentary abstinence on our part is, of course, only one thread in the complex fabric of our alienation, but it is the one I have been exploring in this paper. To remedy this would perhaps be a step toward the sense of identification or kinship with our land that would mark the beginning of a new relationship between ourselves and Nature. And a remedy, at least for this particular mode of our alienation, seems in principle straightforward enough: we initiate ourselves into a native diet. But how is this to be accomplished? Are we to begin *farming* such

foods—cultivating native plants and breeding native animals?
Certainly the farming of native foods would be ecologically
preferable to the farming of European foods, and various experiments on this front are under way.

But before we leave the perhaps rarefied philosophical
questions which have occupied us thus far, and get down to the
practicalities of cultivating, say, wattle seed on a commercial
scale, I have a couple of final philosophical reservations I would
like to register. The first is related to the farming of indigenous
animals (as opposed to plants), and the second is a doubt about
the philosophy of farming per se.

While I have no objections to the sustainable cultivation of
native plants—beyond the general objection to farming that I
am about to raise—I am much more uncomfortable with the
idea of breeding native animals for commercial use. This is a
complex issue, and I do not have time to investigate it here;
suffice it to say that, while the farming of indigenous animals
is undoubtedly preferable, from an ecological point of view, to
running sheep and cattle on marginal lands, it may be dubious
from the viewpoint of animal ethics.

My second objection—to farming per se—returns us to
some of the major philosophical questions which I reviewed very
briefly at the beginning of the paper. Farming—horticulture
and animal husbandry—was an invention of the Neolithic
period of prehistory. This innovation, which is supposed to have
ushered in civilization, also introduced into our thinking the
distinction between the tame (or domesticated) and the wild.
This distinction arguably evolved into the pernicious conceptual
division between humankind and Nature which became the core
of the world view underpinning anthropocentricism. Forager–
hunter peoples of course made no distinction between the tame
and the wild, and accordingly knew no such thing as 'wilderness'. Such peoples took sustenance from the whole of the land:
none of it was out of bounds to them. Accordingly, in their conceptual systems no categorical gulf opened up between humankind and nature: Nature was simply the matrix within which
humanity, along with all other species, made its living.

With the advent of horticulture and animal husbandry, in
contrast, a boundary was drawn between space set aside exclusively for human use, and the space that fell outside these

human preserves. The space outside was thereafter basically beyond the pale of human experience, and inhabited, moreover, by threats to human settlement and human life.[11] I think it is clear how this division between space set aside exclusively for human use and space not in any way utilized by human beings could have provided a template for a conceptual division between humankind and Nature. The reservation of space exclusively for human use sows the germ of the dualistic idea that the theatre for human affairs is and should be removed from the rest of Nature; it also suggests the utilitarian idea that natural systems can and should be replaced by systems serving human ends. Both these ideas may have been innocuous enough in the Neolithic, but as the space allocated for human use has progressively expanded, and the space 'outside' has diminished almost to vanishing point, the tragic consequences of these ideas have become apparent.

This speculation about the origins of dualistic systems of thought sets in motion some large, if at present not very practical, doubts about farming. Does farming, as at present conceptualized, perhaps provide a flawed foundation for human society? Can we devise a livelihood that puts us back *into* Nature rather than fencing us off into separate spaces? Can we collaborate with the spirit of the land, accepting its manifold *gifts* rather than sacrificing it to our superfluous *product*? Why work against the spirit, when the spirit will freely provide for us, when the great labyrinth of energy circuits which is the land itself will prove our best, most reliable and labour-saving 'means of production', and in so doing will simultaneously sustain a multitude of other species? Can we, in other words, envisage an alternative to, or modification of, farming in its present form—a native version of Permaculture perhaps, which combines some of the techniques of cultivation with some of the indigenous techniques of land management? Can we imagine a *modus vivendi* that is more appropriate to the age of ecology? Can we devise a *modus vivendi* closer in spirit to, if technologically more diverse and sophisticated than, the eat-the-land culture of the Aboriginal peoples?

6

Property Rights and the Environment

SUSAN DODDS

Debate about the use of land frequently turns into debate about property rights. Attitudes to property rights may therefore profoundly influence responses to proposals designed to protect the environment. This chapter examines the legal and moral underpinnings of differing theories of property rights to suggest ways in which the values found in them which contribute to environmental degradation may be challenged and reviewed.

Legislators and lobbyists proposing regulation or control of what can be done with private land (for example, increasing controls on the use of certain fertilizers or limiting the amount of land which can be cleared of native vegetation) may be confronted with claims that such control would violate the private property rights of the landowner. However, some of those who seek legislative change which limits legitimate use of land may themselves argue that *their* property rights over their land are infringed by the untrammelled freedom of other landowners to do what they wish with what they own. Landholders who seek increased regulation may argue that their rights to enjoy their land, or to the use of it, or to the value of their land, are diminished by the unrestricted exercise of the property rights of other landowners who may be clear-felling or spreading fertilizers. Alternatively, the state, acting on behalf of its citizens (including its future citizens), may aim to restrict property rights over land (and over factories, storage facilities, etc.) on

47

the grounds that the community has an interest in the conservation of the environment.

These three kinds of response to proposals aimed at protecting the environment (of which there are many other examples) pick out three ways in which values associated with property rights engage with values associated with the environment. The first response could be characterized as a claim that the property rights which constitute ownership are *indefeasible*, which means that no restriction of these rights is legitimate. This raises the conceptual questions of just what property rights are and whether they are absolute moral rights. The second response, that property rights ought to be restricted if their exercise infringes the property rights of others, raises the related *justificatory* questions of what makes distributions of property rights legitimate, and how different exercises of property rights may be valued, given a particular justification of property rights. The third response raises the more general justificatory question of the relative weight of property rights and whether there are conditions under which they may be overridden by other values (e.g. aesthetic values or the intrinsic value of the environment).

To help sort out some of the confusion which discussion of property so often brings to environmental debate, I offer an overview of the conceptual and justificatory features of property rights and their limitation. The philosophic issues touched on here are subject to considerable debate. I can only sketch some of them here. By focusing on property institutions and their justification as they occur in states like ours that have arisen from liberal–democratic principles, I aim to draw out the values on which many attitudes to property and the environment are based. Liberal theories of property place high value on the freedom of individuals to exercise property rights to shape their environment to suit their individual values and projects. The value placed on individual liberty in the exercise of property rights frequently conflicts with other values which might ground restrictions on property rights. Concern for the welfare of future generations or for the intrinsic value of nature provides moral grounds for restricting property rights, hence for restricting some exercises of individual liberty. Examination of the complex relationships among these values, property rights and the environment can assist theorists and policy-makers in determining how property rights and property systems can best be changed to protect environmental values.

The chapter is in three parts. First, conceptual issues about the nature of property rights and ownership are addressed. Then theories of property right justification are considered, and three kinds of property right justification are sketched, showing the values which ground them and assessing their differing responsiveness to environmental concerns. Finally, the ways in which our property institutions could be altered to address environmental problems are considered.

Property, Property Rights and Ownership

Some of the confusion which arises when debate encompasses property and property rights turns out to be grounded in the many ways in which the terms 'property', 'property rights', and 'owner' are used. The word 'property' can be used variously to mean, among other things: the things which one owns (such as when one says, 'This pen is my property'); or the rights which one has over one's property (e.g. 'The income from the lease of my car is my property'); or a system of private property rights (e.g. 'Property is one of the cornerstones of liberalism').[1] The term 'owner' is similarly versatile. Sometimes we use 'owner' to refer to the person who has full property rights over a thing (e.g. 'I am the owner of my pillow, so I can sell it, or lend it to you; I can pull all the stuffing out of it; or use it to prop up my head, etc.'). Sometimes we also use the term 'owner' to refer to whoever has a significant bundle of property rights over a thing, even though other rights over the thing may be held by others (so a mortgagee can claim to be the owner of her farm, even if 90 per cent of the value of the farm is owed to the bank, and the bank has the right to force a sale of the farm if she defaults on her loan). Given these various common usages, it is not surprising that so much confusion is kicked up by prop--erty talk.

While I would not want to argue that any of these ways of talking is unacceptable, it is important to be clear how the various terms are being used in debate. Our interest, here, in property and property rights is generated by concern for legislative and policy change; therefore it is appropriate to develop concepts of property, property rights, ownership and property systems which at least approach those used by legislatures and judges so that the values which underwrite property attitudes may be examined.[2] In legal and political philosophy, the centre of attention in discussion of property is not the thing itself, not the land, house, shares or factory, but the *property rights* which can legitimately be held over the thing. What determines whether or not one is the owner of a thing in a legal system like ours—that is, whether the thing is one's legally protected property—is whether or not one has the appropriate bundle of property rights over it. However, one can have property rights over a thing and still not own it, such as when one has an easement over a piece of land, or when one leases property.[3] In a private property system, then, the *owner* of a thing is the person (or corporation) who holds, legitimately, a sufficient range of property rights over the thing for that thing to be his *private property*.

Property can be understood best as a relation between persons involving *bundles of property rights and responsibilities* rather than as the physical thing over which one has property rights.[4] *Ownership*, according to the legal theorist, A. M. Honoré, is the greatest bundle of these rights and responsibilities which a mature system of law recognizes.[5] According to Honoré, the various incidents[6] of ownership can be enumerated in the following way:

> ownership comprises the right to possess, the right to use, the right to manage, the right to the income of the thing, the right to the capital, the right to security, the rights or incidents of transmissibility and absence of term, the prohibition of harmful use, liability to execution, and the incident of residuarity.[7]

Honoré challenges the view that ownership is a right to unlimited control of what one owns. He argues that if someone held all these rights over a thing, then that person would clearly legally own the thing; but also argues that, in our property system, for instance, many subsets of these rights may be sufficient for an item to be owned. The various incidents of property may be split among a large number of right-holders. No single incident is necessary or sufficient to identify the bearer of that right as an owner of an item, as our laws admit of many different combinations of rights over different items as constituting ownership. Further, as Honoré's list indicates, property rights, in our legal system, may be restricted or limited in various ways without ownership over an item ceasing. These restrictions on property rights occur not only through the general incidents of prohibition against harmful use and liability to execution, but also through such restriction on property rights as various forms of taxation; the power of states, the Commonwealth or local councils to expropriate land for various uses; restrictions on sale, use or disposal of some items (e.g. semi-automatic weapons or heritage-listed properties) and so on. Clearly our existing property system does allow for restriction and limitation of property rights over items.

The prohibition of harmful use limits an owner's rights to use, to management, to the income of the thing and to the capital, by forbidding the thing to be used in ways which cause harm (or can be anticipated to cause harm) to others. For example, most land holders are debarred from using their land to store uncontained hazardous waste, because doing so is likely to harm others. The prohibition of harmful use may be the first place to start in attempts at protecting the environment. However, this incident of property is standardly interpreted rather narrowly as a prohibition against using what one owns in ways which directly harm other existing people or their property rights (e.g. holding factory owners responsible for the noxious fumes or noise emanating from their property). The incident is not usually understood as extending to harm to non-human species or long-term damage to the ecosystem.

Property rights may be limited in various ways. First, a particular property right may be restricted. Two examples of this are capital gains taxes (which limit right to the capital of a thing by restricting the amount of the price of sale that one can enjoy after selling an item), and various planning laws and environmental regulations which restrict the uses to which land or buildings can be put, hence restricting the rights to the use, to the management and to the capital of the thing. Second, a particular item may be such that certain property rights cannot be held over it; hence the range of rights which may be held over the thing are limited. An owner of a piece of land held in trust has no right to sell or transmit the land, but must pass it on to the next beneficiary. A third way in which property rights may be restricted is the division of property rights over a thing among various parties in a way which limits what each can do with the thing: this can make clear identification of an owner very difficult, as in cases of joint ownership or state ownership. In these cases, it makes more sense to talk in terms of various property rights and responsibilities held by various individuals or groups over different things related to the institution in question. Indeed, in law, talk of ownership has virtually disappeared; debate is now over the competing property rights and responsibilities held by various parties and how far those rights and responsibilities do (or ought to) extend, given their impact both on property right holders and on other members of the community.

The point of this discussion of the way property rights are divided and limited in our legal system has been to shed some light on the complex nature of existing property rights and property institutions. Property rights can be understood as any of a number of rights which could be had over a thing: talk about property rights does not presuppose an owner of property, and there is no *conceptual* difficulty in restricting property rights in various ways. What can be difficult, especially with regard to our present legal system, is establishing the legitimate limitation on property rights. To establish the justification for limits to property rights, we need to know what the justification for property rights could be in the first place.

Theories of Justification of Property Rights

Although the property *system* recognized by Australian law can be readily characterized as a private property system[8] grounded in the political theory of liberal democracy, the actual distribution of property rights and their restriction does not clearly reflect a rigorous application of any particular justification of property rights. Someone trying to provide an analysis of the justification of the existing system of property rights and restrictions over mineral deposits, farm land, television stations and private cars might well come to the conclusion that there is no rhyme or reason grounding the law. About the only thing that does seem to permeate the property system is an assumption that property right restriction or limitation must be justified by appeal to the property rights or other legal rights of particular persons or the interests of the community as a whole. However, *which* rights or interests (and, hence, which values) are sufficiently strong to justify limitation of property rights is the subject of much debate. At the heart of this debate are competing theories of justification of property rights.

To argue that certain property rights ought to be limited to protect some aspect of the environment, one needs to show either how that restriction is consistent with the values which ground (or ought to ground) property rights, or that the values protected by the restriction are more fundamental than the right to be restricted. I sketch here simplified accounts of three different *kinds* of property right justification which fit liberal democratic theory. The three kinds of theory of justification can be characterized as those grounded on liberty, those grounded on utility, and plural values.

Theories of property based on appeal to *liberty* start with the assumption that political and legal institutions ought to foster and protect each individual's right of self-determination; that is that each person ought to be free from interference in the exercise of their capacity for self-determination, limited only by the obligation of each to not harm any other person in the exercise of their liberty. Exclusive property rights enhance each individual's liberty by granting to each a private sphere within which they can exercise their liberty without infringing the rights of others. The defender of this justification might offer as an example:

> If I privately own my house, I can do what I please in it and no one else can have any grounds for interfering with what I do in the privacy of my house as long as I do not harm any one else. However, if there were no private property rights, then my capacity to be self-determining would be greatly restricted as I would have no guarantee of protection of the arrangements I ought to be free to make about my life (where I sleep, whether I eat this or that, and how I will use my legitimately acquired land).

Property rights, on this view, are necessary for respect for liberty.

There are at least two versions of the liberal basis for property. I will discuss only the more extreme, libertarian version here.[9] In this account the earth is conceived of as having been initially unowned; the appropriation of bits of the earth and the fruits of the earth by individuals through the exercise of their liberty is legitimate, provided that they do not violate anyone else's rights in the course of their appropriation. Individuals gain full rights of ownership over what they legitimately appropriate—that is, they gain all of the property rights enumerated by Honoré. On this theory, owners are free to do what they want with what they own, limited only by an obligation to compensate others for violation of their rights, or for directly worsening the situation of others through the exercise of property rights.

Because the libertarian justification places so much value on the individual owner's right to freely exercise all the rights of ownership, establishing legitimate restriction of property rights on environmental grounds is particularly difficult. To justify restricting property rights, one would have to establish either that another person's rights were being violated or that their situation had been directly worsened by the exercise of the owner's property rights. For example, a libertarian conception of property might allow that an owner's right to use land in a way which causes pollution to spread over other people's property ought to be restricted, but that restriction may simply amount to a duty to compensate the other property owners for the effects of the pollution on their property, rather than as an obligation to forgo polluting.[10]

The centrality given to the value of individual liberty in the libertarian theory would further preclude regulation of land use to protect against degradation, unless the (current) property of others is damaged by it. Ownership, as libertarians conceive of it, entails no duty to leave what is owned in a useable state when one disposes of it. The most that could be done to reduce land degradation would be to pay landowners some form of compensation in order to purchase their agreement not to use their land in certain ways (a kind of reverse pollution tax). On this justification of property, the onus is always on those who seek restriction of property rights to establish that an exercise of property rights violates others' rights or to provide compensation for restriction of the owner's free use of the property. The libertarian justification of property rights involves a view that the owner's freedom to exercise property rights is a fundamental value which frames all other claims of rights or obligations; as such it is not open to arguments for restricting property rights on the grounds that preservation of an environment or a species is sufficiently important to justify limiting property rights, as the values of species or environments is less fundamental than the property rights of the owner.

Clearly the libertarian version of a liberal justification of property rights leaves little room for regulation of property rights on environmental grounds. A theory of property rights grounded in *utility*, however, involves the rejection of the view that individuality is the fundamental value which grounds legitimate property rights. Those who accept utility-based theories of property would argue that concern for the welfare of all people justifies limiting the freedom of particular individuals to exercise property rights.

Theories of property rights based on utility ascribe property rights by determining which distribution most efficiently provides for the welfare of (all) individuals. Put simply, the liberty of an individual owner to exercise property rights without restriction is seen as less important than the basic welfare interests (e.g. adequate nutrition, health care, clean water and housing) of all people (including future people). It is an empirical question *which* distribution of property rights would best provide for welfare, but there is no theoretical presumption that private ownership of primary resources, such as land, is preferable to state or communal property rights. A property system justified by appeal to utility would involve complicated divisions of property rights over different things. For example, an individual might have exclusive property rights over her toothbrush or bed to use as she sees fit, but might have only limited rights over her house. She may have rights to use, possess and manage, but she may not have a right to use the house indefinitely. She may lack the right to the capital over it, or to its income, because living space may be allocated on the basis of need, or because she does not need the right to sell her house in order to have her welfare interests protected. Land might be deemed to be open to only limited property rights— rights which would allow for restricted use of the land for productive purposes, but which would not allow for indefinite exclusive rights to use the land however an individual might see fit—because the welfare interests of both present and future members of the community demand some access to land and hence would be greatly harmed by both exclusive private ownership or destructive uses of the land.

Clearly, it would be possible to make this kind of justification of property rights based on utility cohere better with environmental concerns than a libertarian justification, at least where concern for the environment intersects with concern for preserving the welfare interests of people. So long as the argument for a particular distribution of property rights and restriction can be shown to enhance human welfare, it can have effect. Hence, at least a 'shallow' ecological ethic could be defended on these grounds.[11] However, there are at least two limitations to this kind of property right justification as a means of achieving environmental ends. First, the current welfare interests of existing people may take precedence over the future welfare interests of future people. Where resources are extremely scarce, it may be clear that preserving for future generations some area of land against use for farming will put a heavy burden on existing members of the community, even though the preservation of that land would benefit some future members of the community greatly.[12] Given that welfare interests are the basis for distribution of property rights and that existing need may be pressing, arguments for some preservationist policies may be overridden by appeal to welfare. The second limitation of this theory for environmental purposes is that, as described here, the welfare interests involved are those of humans. Until human welfare is secured, policies geared to securing environments solely for the sake of non-human species or for the environment itself will be given low priority.

Pluralist theories of property right justification include a number of values as being relevant to the legitimacy of property rights. A pluralist account which is true to liberal democratic principles might include considerations not only of liberty and utility but also of other values such as cultural or historical value, aesthetic value or the intrinsic value of the environment. The *property system* might be grounded primarily in liberty and utility (both the welfare of persons and their individual autonomy being grounds for property right distribution), but other basic values may justify overriding or limiting particular property rights so justified. The property rights justified by a pluralist theory would form a mixed system in which private ownership (in the sense of one person holding the full list of property rights over a thing exclusively) would have limited application, while restricted sets of property rights over many things would be held by the state, communities, small groups, and individuals. Such a justification of property rights could allow for recognition of the legitimacy of restricting property rights out of concern for the preservation of the environment, over and above those restrictions which coincide with human needs or the demands of liberty.

Regulation of property rights by appeal to basic values other than those which ground the property system could occur in cases where recognition of a basic value (e.g. the intrinsic value of the environment) overrides a specific property right which ordinarily would be legitimated by the property system (e.g. the right to manage, or the right to capital). For example, one might be concerned to restrict development of land which currently offers bushwalkers in a nearby park breathtaking views of native bushland, on the grounds that the aesthetic value of the land ought to be preserved. The aesthetic value of the land may well be worth preserving, but there may be other considerations in favour of developing the land, such as pressing need for housing in the area, or concern that the person who holds rights of use and management over the land would be unfairly burdened, and her liberty unfairly restricted, if she were now to discover that she had no right to develop the land. The competing values must be balanced out. Because of the plurality of values on which property right distribution and restriction is based in this kind of theory, establishing the legitimacy of environmental concerns is less difficult, as there is no presumption of the primacy of either the liberty of the owner or human welfare interests. Nonetheless, even in a pluralist theory, environmental concerns will not always be given precedence. Some may see this as a failing of the theory, but I disagree. Concerns for the preservation of the environment may be important, they may be vitally important, but the environment is not the only thing of value which individuals and states ought to be concerned to protect; there are others, such as justice or the claims of indigenous communities. Therefore a balance must be struck in determination of distribution of rights and

responsibilities over things in our environment, one which takes
account of all the relevant values.

I have offered three sketches of property right theories and
have indicated the ways in which property rights and obligations
can be derived from the values which underpin the theories.
As they are characterized here, each allows for certain kinds of
alteration of property institutions in accordance with some environmental concerns, in different ways and to varying degree.
I have also indicated some of the ways in which policies could
be presented for legitimately altering existing property systems
in light of environmental concern.[13]

Conclusion

In Western liberal democracies landowners have been, historically, thought of as exercising a kind of sovereign control over
their land. Any encroachment of the owner's property rights
was seen to threaten the very foundations of political order. No
authority could legitimately restrict owners' rights to do as they
pleased with their land. This attitude towards property rights
and land ownership was grounded in the values of the liberal
theory of property which was dominant from the eighteenth
century.[14] Although legal regulation and limitation of property
rights over land has increased greatly over the past fifty years,
the attitudes of the past have often remained. By recognizing
the ways in which such values are found in particular theories
of the justification of property rights, we can challenge those
attitudes towards nature and property which contribute to the
degradation of our environment. Further, by understanding the
relationships between the range of values protected by different
theories of property and the different values we place on the
environment, we can begin to argue clearly and sensitively for
change in our property institutions. The need to do so has been
underlined by the debates which have occurred in Australia
since 1992, when the High Court in the Mabo case opened up
the question of property and ownership rights by recognizing
that Aboriginal rights to the land may not always have been
extinguished by European settlement.

7

Learning from Other Cultures

KWI-GON KIM

In the past several decades, the development aspirations of Korea have mirrored the lifestyles and economic achievements of developed countries. Very recently, however, Korean people have begun to realize that these lifestyles and economic achievements lead rapidly to environmental degradation. It is now perceived that developed countries are far away from sustainability in the sense defined in Korean traditions. Many people in Korea argue that it is urgent from an environmental standpoint to make our development model more sustainable. There is a growing feeling that, whilst Western scientific and technological approaches have contributed greatly to the modernization of Korea, the undesirable consequences of this development are demonstrating the need for a new, sound approach reflecting the cultural values of Korean society.

The principal aim of this chapter is to improve understanding of the relationships between environmental knowledge, awareness and action connected with growth and development from a Korean perspective, and to suggest prospects for a new relationship leading to environmental sustainability. To explore what this new relationship might be, the chapter begins with a discussion of sustainability and a comparison of Western and Eastern approaches to nature and environment.

Environmentally Sound and Sustainable Development

Environmentally sound and sustainable development (ESSD) has become a task for Korea as for other countries in the world. Nevertheless, sustainability is not a well-defined concept. A judgement about 'environmental sustainability' cannot be made before there is clarification of the nature of the environmental information being considered and the context in which it is being used. The set of elements to be defined is not the same for all environmental systems. In addition, judgements about sustainability must always be made in relation to particular sorts of values. Questions such as sustainability 'for whom' and 'for what' are inseparable from the concept of sustainability itself. This means that the perception of sustainability varies considerably between, as well as within, cultural groups. People who hold different sets of values may choose different actions when faced with the same evidence. Therefore, societal perceptions, attitudes, and values must be considered along with knowledge of environmental systems, as part of the study of transferability of possible courses of action from one culture or subculture to another. Figure 7.1 shows a conceptual relationship between environmental knowledge, values and action.

Faludi argued that there are three paradigms of planning theory and practice.[1] (Here paradigm is defined as a distinctive perspective from which problems and solutions are being approached.) According to Faludi, these three paradigms are characterized by the way they conceive of planning. They can be seen as *object-centred, control-centred* and *decision-centred*. The underlying perspectives on knowledge and action are reflected in these three paradigms.

The object-centred view of planning lays overwhelming emphasis on the object of planning. No attention is paid to the steps and learning from understanding to action. Action simply follows from knowledge (see Figure 7.1). The underlying message concerning knowledge and action is loud and clear: knowledge is objective and certain; knowledge can be obtained through study and is thus possessed by experts (especially by

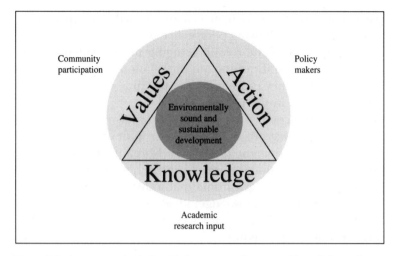

Figure 7.1   A conceptual relationship between environmental knowledge, values and action, and environmentally sound and sustainable development

experts with a special skill in synthesizing such disparate knowledge as there is).

Control-centred planners see planning as state intervention. They are concerned with the nature and effects of powers of control exercised by the state. An underlying assumption is that control powers ultimately serve the maintenance of the capitalist powers of production. Marxists, housing reformers and welfare economists alike hold a control-centred view of planning.

By contrast, the perspective on knowledge and action underlying the decision-centred view of planning is in tune with current epistemology. Existing knowledge forms a tentative framework that has emerged from past negotiations, negotiations in which the researcher committed to searching for truth is but one of the parties to the game. Every argument can come up for renewed discussion. The formation of new knowledge certainly requires fresh arguments. This applies even more to action. There, the relationships between the views held and the goals one pursues are even more overt, and the subjective nature of proposals even more evident.

Rendering the relation between knowledge and action ex- plicit can only make the need for debate and agreement of all participants more evident; it can never make it superfluous. Therefore, the underlying perspective is also a democratic one. Experts have a definite role in bringing in the evidence. They are full partners in thrashing out the relationship between facts and values, but they can never have the final say. They are not the only players in the game.

A decision-centred view of planning has important implications for the way we approach environmentally sound and sustainable development (ESSD). In Korea it suggests that we need to treat traditional Korean approaches to nature and environment, and to grassroots naturalistic philosophies, as important bodies of knowledge, ideas and belief which should be drawn upon and taken into account in environmental planning and decision making.

Historical Perspectives on Western 'Environmentalism'

There are several bases for environmental concern in the West, and they lead to ambiguities in the contemporary use of the term 'environmentalist'. One view holds that nature's resources are there to be used, but they should be used efficiently and without waste. A second view is that the absence of caution in human undertakings can lead to irreversible and sometimes disastrous effects on the ability of natural systems to function. A third view, derived from aesthetic, religious, and ethical concerns, calls for restraint on human actions affecting the environment. Several of these concerns have been translated into ethical norms and in their turn into government policies that guide the way decisions affecting the environment are made. Commonly used norms and policies rest on the idea that human welfare is diminished when natural resources are wasted, air and water is made unhealthy, and so forth. This approach to setting policies and norms is said to be anthropocentric, in that the concern for the natural environment is based ultimately on the welfare of people.

In contemporary Western culture, decisions affecting the natural environment are generally made from an anthropocentric standpoint. However, there are many who argue that an anthropocentric approach to decision-making is too narrow because it does not recognize that non-human species have a right to life, independent of any value to humans.[2]

Naturalistic Philosophies in Korea

A number of underlying factors, including local cultural, religious, historical, and political factors, have affected environmental research, awareness and action in the environmental field in Korea. Because of its location at the far eastern end of the Asian continent bordering China, Korea has traditionally been exposed to Chinese influences in most cultural fields. As well as Buddhism, which has been the main inspiration for many magnificent cultural achievements, the Chinese philosophies of Yin and Yang, the five elements of the universe, P'ungsui, Zen Buddhism, Taoism and Confucianism have been the most evident in Korean society. Being great believers in a life of harmony with nature, Koreans readily accepted these naturalistic philosophies. Among the naturalistic philosophies, this chapter focuses on P'ungsui (geomancy) because it is regarded as a method of land suitability analysis in a modern sense and is a socially rooted determinant of land use in Korea. P'ungsui is an aesthetic science dealing with the positive management of land in accordance with the hidden forces within the earth. According to P'ungsui, a structure was invariably placed to face the south with a mountain at its back. Ideally, the mountain had to have 'wings' at both sides so that it could embrace the structure which, in keeping with Yin–Yang considerations, had to have a stream flowing in the front. Efforts were made to avoid having man-made constructions disrupt the natural contour of the terrain, thereby destroying the highly revered harmony of nature.

An example is shown in Figure 7.2 of a Japanese interpretation of this Chinese naturalistic philosophy in the placement of the city of Kyoto. The city lies on a slightly inclined plane surrounded on three sides by low, rolling mountains. The choice of the site was determined in the eighth century in accordance with the principles of Chinese geomancy. Kyoto's grid plan was a smaller model of the Chinese capital of Ch'ang-an,

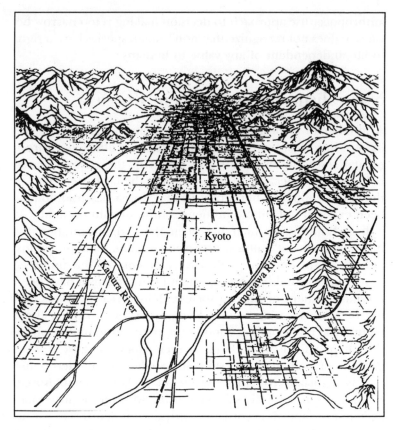

Figure 7.2   Aerial perspective of Kyoto, Japan

and originally measured 5.5 kilometres north to south and 4 kilometres east to west. The north–south axis split the town into symmetrical halves. At the northern terminus lay the imperial compound with its buildings arranged according to the hierarchical offices of government. These buildings, like the entire city, all faced southward, the direction the Emperor always faced, toward light and warmth.[3]

In P'ungsui, people regarded land as an organism with a system symbolizing hidden forces and natural phenomena. Such a system is circulatory. P'ungsui traditions have exerted influence on norms, values, beliefs, perceptions, orientation, and practices related to the environment. Over time the awareness and practices of P'ungsui have undergone changes according to the way in which the dialectical unity between the continuity of purpose to create a better life and the change of form to solve its realization has been grasped and understood. In other words, the interpretation of the P'ungsui tradition has changed during its different phases, even though the tradition has retained its original shape.

Changes in the Influence of P'ungsui

Korea came under P'ungsui influence in the early ninth century. Zen Buddhism was then the main inspiration of daily life. It is believed that Zen Buddhist monks readily understood the theory and technique of P'ungsui because they were familiar with the operation of hidden forces within the earth. They had the political and geographical understanding that the location of Kyongju, the capital of Shilla, was not suitable for the capital of the nation since it was located in the southeast of the nation. Such an idea was accepted by people who dreamed of revolution in the central region, and provided the philosophical and moral basis to the movement led by Wanggun who became the first king of the Koryo period. Zen Buddhism and P'ungsui became the most important ideology of the time.

At the end of the Koryo period, P'ungsui became an anti-government ideology. P'ungsui theoreticians argued that the forces of land no longer existed in Kaesung, the capital city of Koryo. This argument offered a theoretical basis for the relocation of the capital from Kaesung to Seoul under Lee Sung-Gae, the first king of the Chosun period. In this period P'ungsui theory played very positive roles in the construction of palaces and walls, and rearrangements of the administration system of provinces and towns.

With the suppression of Buddhism in favour of Confucianism, P'ungsui became associated with concepts of filial piety in Confucianism (applied, for example, to the choice of burial sites). As a consequence, P'ungsui tended to become an egoistic mediocre religion. P'ungsui theoreticians attempted to exert an influence on government decision-making. Under King Sejong they suggested that the main mountain of Seoul should be the central mountain located to the rear of Seoul. They also suggested that Chunggaechun, a stream flowing in the front of the main mountain, should be protected from water pollution. These ideas were not, however, accepted by the rulers and politicians. Thus, although there was a tradition of thought, namely thought of creating a new world, which supported and made possible the practice of P'ungsui at the end of the Chosun period, P'ungsui theoreticians were no longer influential. For this reason, traditional P'ungsui disappeared from the forefront of politics and became associated with Taoism, emphasizing a life of harmony with nature. P'ungsui was never officially practised during the Japanese occupation.

Modern Environmental Awareness and Practices in Korea

1950–1980

The 1950s were a period of social turmoil in Korea following the Korean War. In the 1960s Korea underwent rapid social changes due to industrialization and urbanization as part of its modernization. During this time the notion of planned national land began to take shape. In a country where the land is restricted and natural resources are scarce, many people argued that the best way to deal with these restrictions was to make better use of the land. Thus, after the Military Revolution in May 1961, the government initiated a series of land surveys, and in 1963 enacted the Comprehensive National Land Construction Planning Law. In this period, the most urgent national task for Korea was to escape from unemployment and poverty. All attention was therefore focused on growth and development policies. Although a certain degree of decline in environmental quality, resulting from government-driven economic development programmes, was observed in the 1960s, environmental deterioration did not receive much attention. Most academics and planning officials were totally preoccupied with economic growth and industrialization. Korea's first environmental statute, the Pollution Prevention Act, was enacted during this period (1963). However, the act was never intended to be an action-forcing statute as no serious environmental problems were perceived at that time.

In 1972 the government enacted the Law on Management of National Land, and restricted the free use of private land. It sought a more efficient utilization of national land, and announced a basic price policy. The government also provided the legal foundation for a water resources development plan and the development of heavy chemical industrial estates through the enactment of the Law on Promotion of Industrial Estates (1973). The First Comprehensive National Physical Development Plan (1972–81) introduced a new stage of planning, but gave relatively little attention to environmental quality. Its basic goals were to improve land utilization and management efficiency, and to expand development.

In the mid-1970s, the environment began to deteriorate and people started to be concerned about the pollution of rivers and coastal waters, and the pollution of air in the major cities and industrial areas which were the centres of growth and development. In response, government relaxed its exclusive concentration on industrialization somewhat, and a new law, the Environmental Conservation Act, was enacted. This act, the first major environmental statute in Korea, passed the National Assembly in December 1977, and was signed by President Park and became effective in July 1978. It required agencies to prepare environmental impact statements and to consult with the Environmental Administration, which had been established as a sub-cabinet agency of the Ministry of Public Health. A new constitution was also adopted in 1980 which guaranteed people the right to live in a clean and healthy environment.

The Environment Conservation Act introduced many important new features of environmental regulation, such as the promulgation of environmental standards, environmental monitoring, setting of emission standards and regulation of emissions, and various administrative sanctions for violations. The act has since been amended several times. Under a 1979 amendment, environmental impact assessment was adopted as a regulatory mechanism. An amendment in 1981 introduced an emission charge system to enforce the measurement of emissions.

In this period the emphasis of Korean planning was on the use of environmental data, models and criteria to maximize the efficiency of land development. Its approach was people-centred and nature-exploiting. Interestingly, the approach was introduced by the young generation educated in Western society after the nation's liberation. This resulted in the weakening of Korean traditions. Traditional Korean attitudes were subordinated to Western scientific and technological approaches to planning. The theory of geomancy had little popular or intellectual currency.

1980s–1993

In 1986 Korea became the twelfth-largest trading nation in the world. Korea was no longer a poor nation struck by war. The people of Korea enjoyed greater economic prosperity than ever before. The 1980s was a period of public pressure leading to increasing governmental interest and intervention to ensure balanced regional development, social and economic quality, and the participation of all citizens in the public-decision-making process. From the mid-1980s, the people and the government of Korea also became more concerned about the environment. Actions to protect environmental quality were largely spurred by central legislation and public concern.

The democratic reforms resulting from the constitutional amendments and presidential election in 1987 permitted environmental groups and concerned citizens more opportunities to express their views. It became obvious that the existing law and administrative arrangements were insufficient to meet these changed circumstances. In 1988 President Roh Tae Woo ordered the establishment of a special research committee to study the capability and performance of existing government organizational arrangements. The committee recommended, among other reforms, that the Environment Administration be elevated to cabinet level. In January 1990 the Ministry of Environment was established, replacing the Environment Administration. Its chief, the Minister of Environment, became a member of cabinet.

In July 1990 the National Assembly passed new environmental statutes which replaced the Environment Conservation Act. These were the Basic Environmental Policy Act, the Air Environment Conservation Act, the Water Environment Conservation Act, the Noise and Vibration Control Act, the Hazardous

Chemical Substance Control Act, and the Environment Pollution Damage Dispute Coordination Act. In March 1991 the Solid Waste Management Act and the Marine Pollution Prevention Act were completely amended. The Natural Environment Conservation Act was passed in 1991. As a result, Korea came to have a complete portfolio of sophisticated environmental laws. The legal and administrative tasks ahead now are how to implement these laws; how to promulgate effectively various environmental policies under the laws; and how to develop laws and regulations further according to changing social and economic circumstances.

The Movement Towards Environmental Sustainability in Korea

There is much evidence that the environmental consciousness which evolved in Korea during the 1980s is now beginning to crystallize into a true societal commitment to achievement of environmental quality. In June 1992 a National Declaration on Environmental Protection was made, heralding the critical need for awareness of environmental issues to become part of everyday life. The declaration states in its preamble that:

> Nature is the mother of human existence. Human beings could not continue to live a day without environmental benefits such as air, water and soil. It is our moral duty to understand human dependency on the environment and to control development within the capacity of the environment.

The main purpose of the declaration is to state that Korean people have a right to a good life and a duty to conserve a pleasant environment. The declaration also sets out the 'Nation's Principles for Conservation of the Environment'. The declaration emphasizes not only the increase in the net welfare value of people, but also ecological and ethical norms for decisions affecting the natural environment. It is expected that these concerns will be translated into government policies, implementation strategies and action plans that will guide the way decisions affecting the environment will be made.

A workshop on Environmentally Sound and Sustainable Development was held in June 1991, sponsored by the United

Nations Development Program, the Economic and Social Commission for Asia and the Pacific, and the Ministry of Environment of Korea, and led by the author. It suggested the following guiding principles to initiate a demonstration of a city environment programme:

1. creating an energy-efficient and imaginative structure of land use and housing;
2. maintaining functions based on self-reliance;
3. creating diversity, natural equilibrium and adaptability;
4. considering human dignity in the urban environmental system;
5. creating cities for future generations;
6. maintaining the relationship between living and non-living things;
7. considering the view of nature in Korea;
8. creating optimal levels of park and open space;
9. utilizing other ecotechniques.[4]

The Third Comprehensive National Physical Development Plan has two aims. First, it encourages eco-development and maintenance of pleasant surroundings. Second, it will create ecopolises—cities in which nature, people and the environment co-exist—based on a comprehensive long-term green plan. This means that the focus of the ten-year national physical development plan of Korea has shifted towards sustainable strategies. The Republic of Korea also signed the Conventions on Climatic Change and Biodiversity during the United Nations Conference on Environment and Development in June 1992.[5]

There is much yet to be done. Despite the above-mentioned social commitment, policies and research works, plans are often overly ambitious or based on inadequate data. In some cases they lack several key components, such as appropriate instruments and personnel for planning, monitoring and evaluation; co-ordination between environmental planning and economic development; and effective channels of communication between national planners and local communications. Nevertheless, it is hoped that such a series of new environmental movements and the changing paradigm for planners will lead to the development of a truly Korean model of environmental sustainability. There is, however, a growing feeling in Korea that this model cannot rely solely on traditional Western means. It must combine the legal, scientific, technological and administrative processes typically employed in the West with traditional Korean approaches to nature and environment.

P'ungsui and Current Environmental Movements in Korea

In the early 1960s Simonds articulated an approach to site planning in harmony with the principles of Oriental geomancy. He stated that:

> The ideally conceived project is one in which the ideal program functions are conceived in awareness of the highest potential of the site and in which both are realized and dramatized together. This ideal site-project expression may be achieved through either harmony or contrast of the project with the forms, planes, or character of the site, or through a combination of both. The project plan and forms must always be determined in full awareness of the total site and its optimum relationship and contribution to the project. Unsuitable site factors are modified, ameliorated, or eliminated. Positive site factors are developed, extended, and accentuated. Project or structure and modified site are conceived together as one. Always there must result a complete unity—a fully satisfying resolution of site and project functions. Such unity is typified by the Chinese symbol yang and yin evolved in the misty beginnings of time and representing the complete and balanced unity of two opposing yet complementary elements—man and woman, earth and sea, and, in planning terms, the functions of the program and the functions of the site.[6]

Over the last decade there has been a revival of Korean traditions in all areas ranging from clothing, food, and housing to Korean outlooks on the world. Houses built of materials used in traditional Korean architecture and embodying its styles are constructed in great numbers. Korean-style houses in areas to be flooded due to dam construction are dismantled and sold to people from urban areas to be reconstructed later. Books on P'ungsui are published in large numbers and have become best-sellers.

In a recent paper, Koh has argued that a choice between 'good development' and 'bad development' should not be based on the value systems of the West, but on the traditional value systems of Korea.[7] In another recent paper, Choi has

contrasted the Dominant Social Paradigm of the West that has
emphasized a utilitarian view of nature with the New Environmental Paradigm that emphasizes respect for nature, to show
how P'ungsui theories could be used to develop a Korean view
of the land in accord with the new environmental paradigm.[8]
The way these approaches differ from the traditional anthro-
pmorphic approaches of the West can be seen from Table 7.1.

**Table 7.1   Comparison of anthropocentric perspective and
P'ungsui theory**

Analytical functions	Anthropocentric perspective	P'ungsui theory
Tenets	The concern for the natural environment is based ultimately on the welfare of people.	The concern for the natural environment is based ultimately on religious and naturalistic philosophic beliefs.
Decision-making	• An anthropocentric basis for decision-making cannot reflect the value of maintaining stable ecological systems or preserving rare species. • The decrease in natural capital due to economic growth and develop-ment is not taken into account • One criterion that is commonly used is an adaptation of utilitarian philosophy known as the benefit-cost criterion.	• A P'ungsui basis for decision-making can reflect ideas of modern ecology and environmentalism. • The essential message of P'ungsui is a life of harmony with nature, in particular with land.
Planning implications	Concern for the natural environment has been translated into government policies that guide the way decisions affecting the environment are made.	Concern for the natural environment has been translated into ethical norms, but not into government policies that guide the way decisions affecting the environment are made.

Analytical functions	Anthropocentric perspective	P'ungsui theory
Planning techniques	• Land suitability analysis. • Carrying capacity analysis. • Plan evaluation and project-impact assessment.	• Cause-effect sequences are not adequately explained by P'ungsui. • The choice of the site is determined according to the organicism concept of land and theory of circulation. • According to the organicism concept of land, heaven, earth and humanity are regarded as one entity. • The basis of Tao is the theory of circulation. The change of day and night, seasonal change and operation of planets are based on the theory of circulation.

The principles of P'ungsui can be seen to have many parallels with new paradigm thinking in science, which now emphasises a holistic and systemic approach to scientific understanding.[9] As knowledge and methodology improve, it might therefore be expected that it will become more and more feasible to make the notion of integration of anthropocentric perspective and P'ungsui operational. Information on hidden forces within the earth may be used to determine the capacity of natural systems to carry a certain size and spatial organization of activities we call urbanization. Presumably such information would suggest areas where priority attention should be placed in alerting planning activists, raising public consciousness, and focusing environmental and land-use controls.

Not only can understanding and applying the principles of P'ungsui help to improve environmental understanding and decision-making; P'ungsui could also be influential in changing the value systems of the general public and particular interest groups.

Conclusions

Korea is now considering how general economic and industrial development policies need to be changed in order to promote simultaneous economic prosperity and ecological stability in the years ahead. The Korean people fully understand that economic prosperity with further environmental degradation is not in the interest of present and future generations.

Humanity cannot survive on this planet without utilizing its natural resources prudently. Every human action affects the world around us to some degree, but the full effect is difficult to assess because of the complex relationships between living and non-living things. Deep understanding of these relationships is essential for an environmentally sound and sustainable development plan. Such understanding varies considerably between, as well as within, cultural groups. The theories of P'ungsui in the East provide the basis of understanding of the hidden forces of nature, which in turn facilitates understanding of the functions and processes of nature. It is, therefore, suggested that the environmental information base should identify both all known environmental forces and their processes emphasized in the West, and hidden forces and their processes within the earth emphasized in this chapter. The combined information should be used by development agencies to identify existing environmental problems requiring remedial action, to gain awareness of the full potential of the environment, and to formulate land-use plans.

A holistic approach to environmental policy dramatically expands the standard explication of the ecological principle that everything in the living environment is related to everything else. If we are going to take seriously the holistic approach, it must incorporate the hidden forces and their processes within the earth emphasized in the East, and consider all links in the chain of environmental events. Actions in one sector of human activity often cause unwanted environmental impacts on another sector, so environmental concerns have to be made an integral part of sectoral policies. This form of linkage expands the environmental agenda enormously.

According to Mann, many observers consider that the achievement of a sustainable society in developing nations is possible only in a system of radical decentralization. It is suggested that, instead of relying on centralized, hierarchical, highly capitalized development, those concerned with development should emphasize relatively small-scale resource use based on community and local experience.[10] In this sense, the grassroots naturalistic beliefs, norms and values at the local level of Korean society must be considered, along with the knowledge of environmental systems, as part of the study of possible courses of action. This new approach assumes a decision-centred approach to planning, as discussed earlier in this chapter. It also requires a dynamic dialectic between the scientific, rational and technological approaches of the West, and the organic and circulatory approaches of the East.

If we are to achieve organic, circulatory, and environmentally sound decision-making leading to the coexistence of people, nature, and environment, the concept of environmental sustainability emphasized in this chapter is considered to be a more useful one than the conventional model. It can be widely used in Korea and in other countries as well.

Part II Knowledge:
Asking the Right Questions

People have been interfering with their surroundings for thousands of years in an effort to improve their well-being. But, as we now know, these surroundings incorporate the life support system provided for us by nature, and this interference often unintentionally damaged the system. No doubt people sometimes noticed the damage, and took steps to ameliorate it, but the historical record is littered with instances of these damaging actions resulting in the collapse of nations and civilizations.[1] There could be two reasons for this. First, the damage occurred so slowly, over hundreds of years, that it was imperceptible, or that the rulers of the day felt that they could neglect it and concentrate on what they saw to be more important. Second, people did not understand the linkages of cause and effect well enough to work out why the damage was occurring, or how to correct it.

The development of science as a basis for the development of society has changed this. First, the power of technology based on science has permitted changes to occur much faster. The great forests of England and Spain were all but destroyed in a couple of generations to build ships. This process has accelerated as technology developed, until today unexpected environmental problems seem to crop up almost overnight (Joni Seager lists some of these in her chapter). Following in its wake has been the insight that damage to our surroundings must have a cause, that the cause is usually to be found in human action, and that, equally, human action can ameliorate it.

So the idea of 'environment' emerged. This word was seldom used in our present-day sense before the 1960s. We did not think about our environment as anything more than our surroundings, much less do anything about it, until we saw that we had problems. But once environmental 'problems' were perceived, governments were called on for 'solutions'. Solutions require improved knowledge; hence environmental research would be needed. Thus 'environment' is an offshoot of the science paradigm. (This idea has already been introduced by Peter Hollingworth in Chapter 4, and will be discussed at some length in the chapters by Joni Seager and Michael Webber in this part of the book.) With the rush of technological development since World War II, and the consequent intensification of environmental problems, the need for environmental action has become part of our conventional wisdom. Hence the push for environmental education programmes, environmental research, environmental policy, environmental planning, and environmental action.

What sort of knowledge do we need in order to know whether action should be taken, and if so, what? Our answer to this question will influence what support should be given to environmental research, and what form it should take. As we saw in Part I, we have conflicting and intertwining environmental values. As a partial recognition of this, it is generally acknowledged that environmental study and environmental research must be interdisciplinary activities. However, the reality, as the papers by Ron Johnston and Tricia Berman and by Brian Finlayson and Tom McMahon both show, is that environmental research as practised in Australia is not interdisciplinary. The funding bodies have not yet worked out how to fund interdisciplinary research, and the researchers have, by and large, not worked out how to engage in it. Michael Webber makes the distinction between what he calls 'physical environmental science . . . an activity devoted to understanding how the natural environment works [and] environmental science, . . . the activity of understanding the physical and social processes that affect environments'. It is physical environmental science that gains nearly all the considerable research funding for environment, as an extension to biology, chemistry, geology, hydrology, meteorology, etc. Johnston and Berman, like Finlayson and McMahon, give examples of interdisciplinary environmental research, but their examples keep within the realms of Webber's physical environmental science, without acknowledging the existence of his larger realm of environmental science. Our everday use of the expression 'the environment' demonstrates how we tend to view our environment as an entity separate from ourselves, an object for dispassionate study by the researcher.

Michael Webber gives a devastating list of pronouncements by various 'environmental' researchers that amount to saying that all governments need to make good environmental policy is more science. Government policy-makers know that this will not do. They have to decide whether to allow a hydroelectric project or a paper mill to proceed, whether to set aside an area for a national park, how to manage the national parks they have already set aside, what advice to give to farmers who are struggling to keep their farms out of the hands of the receivers. Their reaction to the view of environmental research that we described above is to forget the scientists, and go it alone. Finlayson and McMahon complain about this, on the grounds that the policy-makers are then often making decisions on the basis of unfinished research, or worse still, that they are bypassing the professional research workers and setting up pseudo-research programmes of their own. There is an obvious difficulty in the opposition of these two different views of the nature of the world, as both parties are essential players in the solving of environmental problems.

Joni Seager views the problem from a different angle—science, especially natural science, is male-dominated, and is based on a male view of how the world works, a view that tends to leave people out. This matters in all science, but it is critical in environmental science, which cannot just be about how the natural world works, but has got to be about how people live in it. She argues that 'environmental problems are quintessentially social and cultural problems. For this reason alone, we should be leery of assigning science too high a prioriy on the environmental agenda'.

Michael Webber takes this argument further, by arguing it through in the context of land degradation (which is also the environmental problem used in the discussion of environmental action in Part III of this book). He claims 'that science [i.e. physical environmental science] produces knowledge that is inappropriate for solving the problems caused by land degradation'. And physical environmental science is, as we note above, the environmental science which is being supported in Australia. 'Science [he says] is unlikely to figure large in any public debate over remedies for land degradation.' He goes on to point out that we are in even deeper trouble, because our environment is no longer changing only slowly. If we do not hurry, 'there will be no environment to science on'. Scientists would be better advised to investigate conservation than degradation.

Thus, we have two chapters which give excellent accounts of the state of environmental research in Australia and which demonstrate that its base lies in conventional science, and two chapters which claim that this conventional science is only marginal to the real problem of environmental action that environmental research is supposed to support. What becomes clear is that the knowledge provided by the physical and biological sciences is only part of the knowledge that we need, and that in any case we need far more than 'knowledge' before we can take appropriate action. If our growing sense of the wholeness of the world and our concerns about the future are to be reflected in action, we shall also need wisdom and will.

Environmental Research Policy in Australia

RON JOHNSTON AND TRICIA BERMAN

Environmental research, in the sense of research specifically directed to understanding the interaction of humans with their surroundings and providing the basis for ensuring that damage is kept within acceptable limits, is relatively new. It is based on the reasonable assumption that improving the protection of our environment requires the ability to assess the likely consequences of any particular action. Thus we need accurate information on land use, environmental conditions and resources. Intelligent and informed decisions about environmental management require an understanding of the complex ecosystems upon which the future of our productive capacity depends. Because the Australian environment is unique, we need information relating to our own particular situation. Thus a body of environmental knowledge, and the research which produces that knowledge and provides the foundation upon which environment policy should be developed, are both required.

In this chapter we explore four issues that are relevant to environmental research, as to all new research fields contributing, as part of their charter, directly to decision-making: How should the field of research be defined, and where are the boundaries? What data are important? Should the research strategy emphasize focus or dispersion? And how and by whom should parameters be established?

What is Environmental Research?

'There is no single definition of environmental research and the estimated expenditures [on environmental research] have been drawn from many sources, each of which uses its own definition.'[1] The Australian Science and Technology Council (ASTEC) has defined environmental research as: 'Scientific work undertaken to acquire and organize knowledge of natural systems that sustain life in Australia. Such knowledge embraces characteristics of natural systems, any impacts of humans upon them, and measures to mitigate adverse impacts'.[2]

Environmental research policy is a new area. It has to grapple with many questions. What is the appropriate level of funding for research to support sound environmental policy and management? How can priorities be established and applied? How should resources be allocated? And what types of institutional arrangements are appropriate?

It is apparent that addressing issues of environmental research and development (R&D) depends largely on how one defines environmental research.[3] For example, although there is agreement that research on the impact of pesticides on the ecology of lakes and streams can be considered environmental research, there is disagreement as to whether research on the hydrology of lakes and streams should be included in the same category. Similarly, the development of advanced fluidized-bed combustion technologies to burn coal more cleanly falls within the category of environmental technology development. However, should the environmental category contain most or all of the engineering efforts directed at burning coal more cleanly —a goal of virtually all R&D concerned with the combustion of fossil fuels?

In Australia, the concept of ecologically sustainable development (ESD) has been adopted by the government to provide a conceptual framework for integrating economic and environmental objectives. The government has established five general principles as key elements of ESD:
• integrating economic and environmental goals in policies and activities;
• ensuring that environmental assets are appropriately valued;
• providing for equity within and between generations;

- dealing cautiously with risk and irreversibility;
- recognizing the global dimension.[4]

The adoption of the framework for ecologically sustainable development has favoured a whole-systems approach to issues, rather than the traditional disciplinary approach.[5] Interest is moving, for example, from specific environmental issues to a general concern with 'sustainable development'.

To summarize, environmental research is not easily defined. It is a relatively new area of research, and is regarded by those working in traditional disciplines as being marginal. It is characterized by a focus on complex, interdisciplinary problems. It is constantly being redefined as knowledge and techniques advance, and as the significance of issues to be addressed changes.

Establishing Sound Environmental Data

> It is not possible with the present form of reporting to determine how much money is being spent on salinity/waterlogging research in the Murray–Darling Basin. Government departments, statutory authorities and universities do not keep such a separate accounting of the multifaceted work funded by their own budgets.[6]

There is currently no adequate database of environmental research funding in Australia, because, as noted above, there is no agreement on what constitutes environmental research, the statistical parameters for categorizing environmental research are still being developed, and there is no requirement that research projects be registered nationally.

There have been attempts in other countries to organize environmental research data systematically. Thus, the UK Cabinet Office Advisory Council on Science and Technology has categorized environmental research according to region of study, principal discipline, and the medium (solid earth, seas, built environment etc.).[7] On this basis it is possible to identify some important characteristics of the research effort. For example, funding for local environmental research is significantly less than that for global and regional R&D. Physics dominates in terms of discipline, and terrestrial and atmospheric research dominate in the medium.

Such a breakdown of Australian environmental research is not possible at present. The Australian Bureau of Statistics has released a compendium titled *Australia's Environment: Issues and Facts*. It clearly demonstrates the problems with existing data sources; the value of attempting to adhere to standard frameworks, classifications and data items; the inconsistencies in current data; and gaps in current data sources.

'Labelling' of environmental research and the establishment of national environmental indicators are crucial tasks for appropriate planning and review of policy directions. It also appears important that those who make policy and funding decisions should have an input into the appropriateness of 'labels'. Thus, a major objective of environmental research policy must be to provide mechanisms for communication and negotiation between the research and the decision-making communities, in order to establish interim agreements on adequate definitions which can subsequently be continuously and purposefully refined.

There have been many expressions of the inadequacy of currently available environmental data. The Resource Assessment Commission highlighted the lack of adequate information on environmental change in Australia, and of definitive information on the causes of such changes.[8] The *National Strategy for the Conservation of Australia's Biological Diversity* has noted that the lack of knowledge of Australian biodiversity, at the level of the ecosystem, the species and the gene, is a serious impediment to the planning and management of its conservation.[9] Australia's Economic Planning Advisory Council stated that 'inappropriate decisions can occur unless decision makers bear all the costs and benefits of their decision, including environmental and ecological implications and intergenerational trade-offs, and decision makers have access to appropriate information'.[10] The case study on salinity and waterlogging in the Murray–Darling Basin compiled for ASTEC called for large-scale research into the effects of salinity on crops, native vegetation and aquatic life, saying much of the advice being given to landholders had a very small research base and still depended on overseas research results.[11]

There is also strong concern about Australia's inadequate baseline data. Long-term monitoring data are needed to indicate trends and patterns of change. Data on ecosystem processes, for example, are needed to achieve sustainable management of natural resources. Examples of sustainable management of resources which rely on long periods for their validity include the following.

1. *Slow processes* such as succession, soil development, and population dynamics in long-lived species where generation time is more important than calendar time.
2. *Rare events* such as reproductive patterns in stressed communities, flood and drought events, and fire events.
3. *Processes with high variability* such as rainfall-driven events in terrestrial and aquatic systems.
4. *Subtle processes* such as those where year-to-year variance is large in comparison with any trend. These need long-term studies to separate pattern from noise. Examples include acid rain and nutrient losses from catchments.
5. *Complex phenomena* involving interacting factors that require large data sets to permit multivariate analysis. An example is population dynamics in aquatic systems.[12]

Impediments to long-term environmental monitoring include the existing annual budgetary cycle, competitive granting schemes, demands for cost recovery, and political pressure for quick solutions. However, there have been some successes in the area of data collection. The *National Geological Mapping Accord* is an agreement between the Commonwealth and states for long-term sustained data collection to support resource development, land-use decision-making and environmental protection. It has established priorities for data collection and is producing digitized geological maps of Australia. As the maps are being compiled digitally, data can be easily manipulated and analysed and interwoven with other data for making decisions.

Another data issue concerns access. Access is about locating data and establishing standards for collecting, sharing and exchanging data. The Commonwealth government has addressed the issue of access to environmental data and information by the following measures.

1. Creating the National Resource Information Centre and the Environmental Resources Information Network. These bodies are negotiating with states to provide directories for

locating data, and they will also enable effective analysis of sets of data through the use of geographic information system technologies.

2. Establishing a new Commonwealth Spatial Data Committee to address integration of spatial data through common standards and co-ordination.

3. Strengthening existing co-operative organizations, such as the Australian and New Zealand Land Information Council, which examines land information data policy.

An Appropriate Environmental Research Strategy

There are two approaches to the organization of environmental research, not necessarily exclusive. The first is to include an environmental aspect in a wide range of research fields. The second rests on a view that some aspects of environmental research, such as measuring or monitoring, require dedicated programmes and institutions.[13]

The first of these, the 'dispersed' approach, has the objective of ensuring that environmental objectives become part of overall R&D thinking and results. However, the integration of environmental research into the established R&D systems may be difficult because of institutional constraints. These include the difficulty of forming interdisciplinary teams and programmes; the relatively low status accorded to the (until now) marginal area of environmental research; and the disadvantage of environmental research in the area of funding (because of the immaturity of the field), particularly in relation to basic research.

Institutions dedicated to environmental research (e.g. the Australian Institute of Marine Science or AIMS) are invaluable for certain types of activity, such as monitoring system properties and impacts and, perhaps, technology assessment. They are less likely to be subject to pressures from other demands, and can better focus on environmental issues; they are, however, constrained by the conditions which define their areas of activity. In addition, at least outside the universities, they must retain their credibility as impartial sources of findings and avoid charges of becoming lobbies; this becomes more difficult when funding comes from a single source.

Four types of framework have been adopted in OECD countries to manage the funding of environmental research:

- a Ministry of Research and Technology that includes responsibility for environmental research;
- an Environment Ministry where there is no clearly defined Ministry for Research and Technology;
- a Research Council dedicated to environmental research (as in the UK);
- an agency with a market orientation towards technological development.

In Australia, the Commonwealth government exercises its responsibility for the environment largely indirectly, by using its international obligations and powers, along with powers related to foreign investment, defence and nationhood. Environmental policy, therefore, does not fit neatly into any of the above categories. Recently it has been substantially integrated throughout all portfolios largely as a result of the adoption of ESD. However, day-to-day responsibility for environmental matters rests with the states.

Funds for environmental research are provided by a variety of portfolios within both Commonwealth and state governments. For example, in the Commonwealth government, the Australian Research Council funds competitive research and is located within the Department of Employment, Education and Training. The Department of the Arts, Sport, the Environment and Territories funds some research with more specific objectives: research linked to greenhouse and climate change, and more recently, the directed research of the Commonwealth Environment Protection Agency. Other portfolios also fund environmental research, some of it basic and some directed to a particular end—e.g. Industry, Technology and Commerce (CSIRO, AIMS, ANSTO); Defence (DSTO); Health (NHMRC); Primary Industries and Energy (BMR, and BRR).[14] The industry R&D corporations also fund research for specific industry sectors, but there is an element of environment protection and sustainable development within them. Non-government funding of environmental research is small, and includes that associated with environmental impact assessment.

The adoption of ESD by the Commonwealth government has involved a major consultation process with public and private sectors to establish strategies for managing change. This process has enabled strong integration of environment issues, including research across industry sectors and between government departments. It has been the major driving force in the integration of environmental values across the community in the past two years.

The Murray–Darling Basin Commission provides an example of co-operative management of environmental problems in a particular natural system. Its Ministerial Council has developed a *Natural Resources Management Strategy*, which is an integrated framework to achieve objectives for the basin as a whole, involving Commonwealth and state governments and local agencies and community groups. The Ministerial Council has achieved unified action from the states on the issue of salinity control in the River Murray.

Another example is the Great Barrier Reef Marine Park Authority (GBRMPA), which is responsible for management of the reef. The authority and the Queensland Department of Environment and Heritage have set up Marine and National Parks over most of the World Heritage Area, and other agencies have implemented a variety of management regimes to protect the area. The authority oversees research (water quality, effects of fishing and crown-of-thorns starfish) as well as day-to-day management of the area (long-term monitoring and research into socio-economic questions concerning the Great Barrier Reef). GBRMPA encourages regional co-operation and co-ordination between those who use and manage the area—for example, institutional collaboration by inviting joint submissions for its contracts, and development of educational and consultancy arrangements. The latter involve James Cook University, AIMS, and GBRMPA, and are aimed at marketing the tropical marine scientific and management expertise and programmes of all three agencies, both nationally and internationally.

Obviously, there needs to be a fine balance between the integration of environmental R&D across all sectors and government, and the management of R&D for effective outcomes. Trade-offs and compromises must be made between focused and dispersed systems; small individual research and large interdisciplinary programmes; short-term research designed to support immediate regulatory needs and long-term and anticipatory research directed at preventing and mitigating future problems; and R&D focused on national concerns versus those efforts directed to international and global concerns.

The Establishment of Environmental Research Priorities

> As Australians, we need to recognise that we constitute less than 0.3% of the world's population yet have stewardship of 5% of the world's land area and a similar area of continental shelf. The money we can spend on understanding our natural environment and its endowment of resources is spread very thinly and must be spent to best effect. This is a challenge for all of us.[15]

An important challenge for environment policy-making is to effectively link the research community with the policy-making community in establishing research priorities. Here, two fundamental forces are often working against each other. First, scientists require independence and continuity of funding to undertake sustained investigation which they believe is challenging, will yield results, and will advance their careers. The second force, which may, and often does, oppose the first, is the need for national agencies to develop the data and the knowledge base that will allow them to pursue their missions and objectives as defined by the government.

To operate effectively, national R&D programmes must be organized, and must operate, in such a way as to achieve a balance between their independence and continuity, and their mission-orientation and relevance to policy. To achieve this balance may require the creation of an organizational buffer (such as the ESD consultation process) that directs and modifies the flow of information from the scientists to the policy-maker and vice versa.

ASTEC reported that environmental management in Australia at both Commonwealth and state levels is fragmented between a range of departments with quite different mandates.[16] It argued that the Commonwealth government, as the major supporter of environmental research, should adopt a framework within its own administrative organizations which allows for effective co-ordination of the many participating interests without removing their primary responsibility. Such a framework would include determination of environmental research priorities. A similar recommendation was made in the ESD Intersectoral Report recommendation for a national ESD research strategy.[17]

In this context, it is worth mentioning the Australian Academy of Science report identifying priority research for global change in Australia over the period 1992–96.[18] The report is the outcome of a number of workshops with more than six hundred participants. The ten major research projects identified tended to focus on interdisciplinary research combining the physical and chemical sciences with biological sciences. Such priorities developed by the science and technology community provide a useful basis for discussion with and consideration by policy-makers.

Conclusion

Environmental research policy is in its infancy in Australia. Given the great importance and variety of the problems to be addressed, and the advantages of an improved understanding of the requirements for effective research management, the challenge is to ensure maximum co-ordination and effective and responsive direction-setting, in order to provide the basis for making informed policies and decisions.

9

Funding and Conduct of Environmental Research

BRIAN FINLAYSON AND TOM McMAHON

An earlier version of this chapter appeared in Search, *vol. 24, no. 5, 1993, pp. 120–4.*

Australia's environmental problems have two principal causes. First, the intellectual roots of Australian environmental science (and of European Australian culture generally) lie in northwest Europe and North America. Our science is based on a conceptual model of the natural environment and of human interactions with it that are not appropriate for this continent. Early agricultural practices, too, were inappropriate to the Australian landscape. Climatically and hydrologically Australia is more variable than the northern hemisphere regions where many of our concepts were developed.[1] Second, Australia was colonized by a centralized bureaucracy, and our political and social institutions still reflect this reliance on 'the government' to solve problems. Government and quasi-government organizations dominate the environmental research arena, at least numerically.

In our review of the funding and conduct of environmental research in Australia, we will argue that the strangeness of the environment and the stranglehold which government bureaucracies have over research funding limit the nation's capacity to cope with the detrimental impact of European-style economic development on the natural environment.

We begin by defining our terms. What is research and how is it distinguished from related data-gathering activities? We review the present situation in terms of who does environmental research and how it is funded. As described by Johnston and Berman in the previous chapter, the Australian Science and Technology Council (ASTEC) has recently attempted a full audit of environmental research spending in Australia.[2] Rather than go over the same ground, we provide here a view from the working level of environmental research. Finally, we draw attention to the limitations of the present funding system and suggest an alternative model.

Defining Research

Research, as a term, is in danger of becoming devalued and losing its precision. It is currently being used to describe a whole range of activities which do not properly fall within its scope. The dictionary meaning of research is 'diligent enquiry or examination in seeking facts or principles; laborious or continued search after truth'.[3] Note the specific reference here to principles and to truth. In its recent study of environmental research in Australia, ASTEC adopted the following definition of environmental research: 'Work undertaken to acquire and to organise knowledge of natural systems that sustain life in Australia. Such knowledge embraces characteristics of natural systems, any impacts of human actions upon them, and measures to mitigate adverse impacts.'[4] Note that, in this definition, the word 'knowledge' means more than just 'data' or 'information'.

A scale of activities can be established to cover the range often now encompassed by the term 'research'.

- *Monitoring* is the routine collection of data. In the environmental field this includes activities such as the gauging of river flow; measurement of contaminant levels in water, air or soil; the collection of data by remote sensing (air photos, satellite); and counting the number of invertebrate taxa present on a streambed.
- *Investigation* refers to the application of a routine methodology to a specific problem. Investigation often draws on the information collected by monitoring. It does not develop or improve fundamental understanding.

- *Research* may arise from investigation if the issues are pursued in depth. Research is also generated by the curiosity of individuals; by the identification of problems which need to be solved; and by the need for a better understanding of the systems we are required to manage. Research is distinguished from investigation by challenging the fundamental underlying concepts which apply to the issue. In a more practical sense, research is characterized by a set of conventions which require it to be published or communicated in such a way as to be exposed to the scrutiny and criticisms of the research peer group and the wider community. Learned journals nurture this tradition of research through the peer review system and the practice of publishing comments and discussion about papers and replies by authors. This process of communication and review is central to the research process.
- *Scholarship* involves taking research into the wider intellectual arena through the publication of learned discourses, textbooks and reviews. Together, research and scholarship add to the store of knowledge, develop new ways of looking at issues by challenging and reformulating the conceptual basis from which they are addressed, and, occasionally, may develop a new paradigm. In good research and scholarship, nothing is sacred and nothing is taken for granted.

It is of fundamental importance that environmental research be seen as an activity that progresses beyond monitoring and investigation, and that research and scholarship, as defined here, are pursued. While we have avoided starting a debate on the meaning of 'environmental', we do wish to make one point: environmental research is interdisciplinary in nature. It follows that funding arrangements for environmental research need to be so organized as to actively facilitate truly interdisciplinary, co-operative activities. As an illustration of this interdisciplinary aspect, consider a study of the sex habits of the mayfly in a river. While this has environmental relevance, it is strictly a study within aquatic ecology or entomology. An analysis of the flow-duration characteristics of the same river is a hydrological study. These two studies, pursued together, constitute an interdisciplinary study of the mayfly breeding habitat, and it is this which we would wish to call environmental research.

Funding of Environmental Research in Australia

What passes for environmental research in Australia, whether it fits the definition stated above or not, is carried out in the following types of organization: government departments (state or Commonwealth); government instrumentalities (e.g. Murray–Darling Basin Commission, Rural Water Commission); the Commonwealth Scientific and Industrial Research Organisation (CSIRO); universities; private companies and foundations; and community groups and individuals (e.g. Landcare groups).

The traditions of dependence on government which have characterized Australian society since 1788 have ensured that little environmental research is carried out here by private organizations. Similarly, environmental research in Australia is largely government-funded. ASTEC drily comments that industry's contribution 'is not easily identified but can be assumed to reflect the relatively low level of industry funded research and development in this country'.[5] ASTEC puts this 'relatively low level' at 1.3 per cent of total environmental research funding over the period 1984–87.[6]

If we use the ASTEC data as a guide, total spending on environmental research was $222.7 million in 1986–87.[7] The Commonwealth government provided $187.3 million, and the state governments $30.2 million, with the remaining $5.2 million from various private sources. The Commonwealth government's share was distributed as follows: $65.0 million to CSIRO, $66.6 million as operating grants to universities, and the remainder, $55.7 million, either spent in-house by Commonwealth departments or distributed through research and development granting programmes. Distributions by some of the more important granting programmes are discussed further below.

It appears that the bulk of state government funds for environmental research is spent in-house in departments and agencies, though data are difficult to acquire. In Victoria, for example, there is no central repository from which data on funding of environmental research can be obtained. A profile of funding would have to be assembled from individual departments. We have been able to acquire details of spending on environmental research for the Department of Conservation and Environment for 1991–92, which shows that that Department alone spent $7.1 million, of which $1.5 million was provided by external sources, mainly Commonwealth grants.

The Australian Research Council (ARC) is the primary source of funds for non-medical university research. Funds are distributed through a competitive system based on peer review of applications. There are twelve review panels, which cover the areas of earth sciences, biological sciences (molecular cell biology), biological sciences (plant and animal biology), chemical sciences, engineering, engineering and instrumentation, humanities, physical sciences, social sciences, cognitive sciences, materials science and minerals processing, and Australia's Asian context. The ARC concentrates on pure and strategic research and rarely commits funds to a project for more than three years. Over the period 1980 to 1989, approximately 30 per cent of its funds were spent on environmental research each year, and in 1989 this was $16.6 million.[8] University researchers pursuing curiosity-driven research seek funds through the ARC. It is important to note that the mechanisms used by the ARC to allocate funds do not encourage interdisciplinary research. Single-discipline applications from researchers who are already well recognized in their fields are those most likely to receive funding.

The Land and Water Resources Research and Development Corporation has recently been formed to take over the research funding responsibilities of the Australian Water Research Advisory Council and the National Soil Conservation Program. The corporation has a strong emphasis on interdisciplinary research and stresses that it supports research that can be applied readily by resource managers. Favoured areas for research funding are clearly specified in the annual call for applications, and considerable emphasis is placed on co-operative applications between researchers and resource management agencies. These funds are available to universities, CSIRO and government departments and agencies.

The Murray–Darling Basin Commission distributes funds competitively to researchers in universities, CSIRO and government departments and instrumentalities through its *Natural Resources Management Strategy*. In 1989–90, the total amount spent was $5.05 million. The priority areas for funding are carefully specified by the commission to address management problems. This programme lacks a rigorous merit-based assessment procedure, and much of the decision-making is done by the commission on what appear to be other than purely scientific grounds.

A variety of other Commonwealth-funded granting bodies, such as the Fisheries Research and Development Corporation and the Office of the Supervising Scientist for the Alligator Rivers Region, also support research in the environmental area.

Limitations of the Present System

It is clear that a sizeable proportion of the environmental research budget in Australia is spent by and within Commonwealth and state government departments and agencies. From the ASTEC data,[9] we estimate this to be at least one-third, but the proportion is probably larger. There are a number of problems associated with research performance in this area. Within many of these departments, there is no clear distinction between research priorities and outcomes, and political decision-making. Departmental researchers are located within the chain of command which leads to the minister. In these circumstances, the independence of research decision-making from the political decision-making process cannot be either demonstrated or guaranteed. The political time-frame also intrudes into the research time-frame. This can lead to a situation where results are demanded before the scientific process is complete or, alternatively, potentially useful research is terminated in midstream because the political agenda has moved on and the resources are diverted to serve some new purpose.

Overall, the staff in government departments are poorly qualified to design and conduct research programmes. Instances of persons untrained in research rising from positions as technical officers to lead research teams are all too common. This situation is exacerbated by poor staff development practices in many departments. Research staff generally are not encouraged to upgrade their qualifications or to participate in research conferences and workshops.

A fundamental requirement for scientific research, as outlined earlier, is that the results be exposed to criticism, comment and peer review. In most of the government departments
this element of research is almost entirely absent. Consider, for
example, the number of papers on salinity research in Victoria
which have been published in the refereed journals (a common
measure of productivity in research) in relation to the size of
the budget so far expended on this exercise. Table 9.1 lists the
research budget for the salinity programme for the financial
years 1987–88 and 1988–89 and the numbers of papers published in refereed journals. The equivalent data are also given
for the Department of Geography, University of Melbourne, of
which the first author is a member. The university data are for
calendar years. The budget of the university department is not
just the research component, but the total cost of providing
undergraduate teaching, postgraduate training and supervision,
infrastructure support and administration. The data in the table
do not include scholarly works such as books and chapters in
books or publications in conference proceedings. If these had
been included, the differences between the university department and the salinity programme would have been even greater.

Table 9.1   Comparison of funding and research productivity of
sample government and university programmes, as measured by
papers published in refereed journals

Funding unit	Year	Annual budget (million $)	Published papers	Cost per paper (thousand $)
Victorian salinity	1987–88	6.5	7	930
programme	1988–89	7.3	11	670
Geography Department,	1988	0.8	13	60
University of Melbourne	1989	0.9	23	43

Source: Salinity Bureau, *Victorian Salinity Program: First Annual Review, 1987–1988*,
and *Second Annual Review, 1988–1989*, Department of Premier and Cabinet,
Melbourne, 1988 and 1989; University of Melbourne, *Research Report 1988* and
1989, Parkville, Victoria, 1988 and 1989.

Earlier in our paper we established distinctions between monitoring, investigation, research and scholarship. Much of what passes for research in government agencies, and is reported in funding summaries as such, is really monitoring and investigation. We can continue to use salinity research in Victoria as an illustration. It is an article of faith among government salinity researchers that dryland salinity is a response to alterations to the rate of groundwater recharge as a result of the clearing of native eucalypt forests and their replacement with pasture and crops. Management plans are now being prepared across Victoria to cope with this problem, despite the fact that no properly conducted and fully documented water-balance study has ever been published for a dryland salinity site in Victoria.

Work by McMahon and Finlayson gives an example of the possible errors.[10] Table 9.2 shows the results of the application of four different modelling strategies to the Axe Creek catchment near Bendigo in north central Victoria. The models become progressively more complex. Model 1 is a simple water-balance equation. Model 2 is the surface rainfall-runoff model HYDROLOG, where groundwater recharge is the deep seepage output of the model.[11] Model 3 is an improved version of HYDROLOG, known as MODHYDROLOG.[12] Model 4 is an integrated surface-groundwater model which simulates appropriately both surface and groundwater components and preserves the interface between them.[13] Note the wide range of possible groundwater recharges, depending on the model used. Given the present state of knowledge, Model 4 is the best alternative, yet to date it has been used only once, and then in a university postgraduate research project, and not in the development of a salinity management plan.

Political pressure to produce management plans has stopped the research process at a point well short of a full understanding of the problem. In the long term, we will need this full understanding to manage the problem properly, but meanwhile the opportunity seems to have been lost.

A further criticism we have of this type of government research is the tendency for old material to be continually reworked and requoted until it achieves the status of fact simply by being quoted often enough. The Avon River in Gippsland,

Table 9.2 Estimates of groundwater recharge at Axe Creek in north-central Victoria, using four different modelling procedures (all values are in mm)

Year	Precipi-tation	Runoff	Pan evap'n*	Model 1	Ground water recharge Model 2	Model 3	Model 4
1981	722	165	419	138	90	67	90
1982	213	0	344	–131	2	2	1
1983	756	138	403	215	103	79	45
1984	607	38	440	129	52	40	11
1985	574	21	441	112	61	47	15
1986	656	105	473	78	107	83	25
1987	664	76	443	145	79	60	24
Avge	599	78	423	98	71	54	30

* Pan evaporation values were used only in Model 1. In the other models, potential evapotranspiration estimates were used.
Source: T. A. McMahon and B. L. Finlayson, 'Australian surface and ground-water hydrology: regional characteristics and implications' in J. J. Pigram and B. P. Hooper (eds), *Water Allocation for the Environment*, Centre for Water Policy Research, University of New England, Armidale, New South Wales, 1992.

Victoria, is a good example. The Land Conservation Council ascribes the recent channel changes on the Avon to changes in the land use in the catchment, despite the fact that the catchment of the Avon (as distinct from its floodplain) is still covered by virgin forest.[14] The popular beliefs that the Snowy River now delivers more sediment to the floodplain than it did before European settlement, and that its channel has silted up, play an important role in generating management activities there, yet it has been shown by careful analysis of the historical data that this is not the case.[15]

The data given earlier imply that a sizeable amount of money is being spent on environmental research through the ARC and the CSIRO. However, this needs to be interpreted with caution, especially noting our earlier comments regarding inter-disciplinary research. The panel structure of ARC and its emphasis on excellence and peer review predetermine that the bulk of the funds go to established specialists in clearly defined disciplines. It is almost impossible to obtain money through the ARC funding mechanism for interdisciplinary research. It is difficult to find a panel to accept an application which is not clearly in their defined disciplines, and referees are reluctant to report favourably on interdisciplinary project applications.

Recent changes to the funding arrangements for both universities and the CSIRO are tending to constrain their activities more into investigation than research. The CSIRO is now required to generate 30 per cent of its income from industry, thus severely constraining its ability to conduct basic research or to maintain long-term programmes. Changes to university funding arrangements, discussed further below, have also forced the universities to become more oriented towards applied contract research.

The ARC, as it exists at present, was formed in 1988 to replace the Australian Research Grants Committee. Over the period 1988 to 1992, the total funds available to the ARC to distribute to researchers in universities rose from $79.5 million to $208 million in constant 1989 dollars. Larkins, while indicating that the additional funds have been used to address some serious deficiencies in the system, points out that this increase is not all new money and has been achieved at some cost to the universities, particularly those which were in existence before 1987.[16] Some of the additional funds which the ARC has had at its disposal have come from the 'ARC clawback', whereby discretionary research funds, which formerly were paid directly to the universities, are now distributed competitively via the ARC. As a result, these institutions now have reduced ability to influence their own research agendas directly, and they cannot adequately support young staff or promote new initiatives in unfashionable research areas. One of the biggest losers under these new funding arrangements has been curiosity-driven research. In these conditions, interdisciplinary environmental research has suffered badly, as the emphasis in research has followed the money into the types of research most likely to succeed in the annual ARC funding round.

An added complication for the universities is that all this is occurring at a time when they are experiencing a shortfall in funding for research infrastructure. A survey commissioned recently by the Australian Vice-Chancellors' Committee has shown that there is inadequate funding to the universities for maintenance, repair and upgrading of equipment and for the maintenance of research collections in the libraries.[17]

Finally, we need to consider the limitations of the present system with regard to training for research. Despite the large spending on research, virtually no training is carried out in the government agencies or the CSIRO; the universities are supposed to provide the next generation of trained researchers, but are not being adequately funded to do so. The most common funding source for graduate students is the Australian Postgraduate Research Award scheme and a smaller number of inhouse scholarships provided by some universities (for example, the Melbourne University Postgraduate Scheme). In environmental research, industry-funded postgraduate places are less common than in disciplines with a strong industry base. Students receive Postgraduate Research Awards on the basis of their proven performance as undergraduates, and, while they are not conditional on the students following any particular research project, the nature of the research the student chooses to pursue will influence the dollar amount of the scholarship received. However, a living allowance is all the student receives from the system; research costs must be borne by the student's department or come out of the student's own pocket. As a direct result of the 'ARC clawback' discussed above, most university departments now find it extremely difficult to provide the financial support needed for high-quality postgraduate research in environmental topics, which usually require field work as well as laboratory and data analysis facilities.

An Alternative Approach to Environmental Research Funding

We identify below characteristics which we see as desirable in funding environmental research, following which we attempt to formulate a new approach.

- A range of funding types needs to be available to support all the categories of activity we identified earlier—monitoring, investigation, research and scholarship. In the areas of research and scholarship there should be support for both curiosity-driven research and applied research. There also need to be mechanisms which fund research by individuals, as well as providing for the establishment and maintenance of research groups.

- Funds are needed to support and encourage research that succeeds in bringing researchers together from diverse discipline backgrounds. This is not a trivial task. Decades of lip-service to the idea of interdisciplinary research have so far not succeeded in generating much activity.
- Better arrangements are needed for supporting research training, both pre-vocational and in-service. It needs to be recognized that the best environment for training researchers is in pure research, where the right mind-set, that critical questioning of all the fundamental assumptions, can be developed. For in-service training, it is important that researchers in government departments and agencies move out of their organizations to interact with groups of professional researchers, for example in the universities or the CSIRO.
- In competitive research funding through government granting programmes (other than the ARC), it is necessary to establish mechanisms which prevent potential recipients from also making the allocation decisions.
- In the public service, there needs to be a clear separation between those involved in the political process (for example, giving advice to ministers) and those involved in scientific research. This 'scientific civil service' needs to be provided with a career structure based on research performance and not seniority or political favour.
- Those research-funding programmes which have a brief to look at particular areas need to develop a strategic agenda for setting research priorities. An example of how this has been tackled already can be found in the *Natural Resources Management Strategy* developed by the Murray–Darling Basin Commission.[18]

A constraint on the new approach we attempt to develop is that it must involve as little change as possible to the present system, consistent with the requirements outlined above. Dramatic changes have little chance of being implemented, desirable though they may be.

Curiosity-driven environmental research can best be fostered by reinstating more direct research funding to the universities. Money is distributed more efficiently this way, since there is no expensive bureaucracy required to organize a competitive funding programme. This money should be channelled straight into the budgets of academic departments, where it will be directly available to researchers. We note that this point has been made quite independently by Jackson.[19]

Another panel should be established in the ARC which deals only with interdisciplinary environmental research.

There are at present four ways in which government departments fund research: by employing research groups in-house; by combining in-house research with some contracted-out research; by setting up semi-autonomous research institutes; or by setting the research agenda and then letting contracts to outside research groups to carry out the work. In the environmental research area the first examples of the institute structure are beginning to appear, and we see this as a positive development. Of the other three, we would advocate that there be no enclaves of researchers in government departments and that these departments contract out their research to professional research groups. The process proposed here should not be confused with consulting contracts.

For the agencies providing research funding, there is a need to develop a strategic agenda, as discussed above. In pursuing this agenda, they need to promote a range of funding types. Before it was disbanded, the Australian Water Research Advisory Council had developed just such a range of funding types, which were well regarded by the research community. They consisted of a National Priorities Programme, a Partnership Research Programme, a Scientific Merit Programme, Postgraduate Research Awards, Research Fellowships, and Eminent Researcher Fellowships.

The Co-operative Research Centre programme now being established by the Commonwealth government has the capacity to meet some of the needs we have identified. At present, five out of thirty-five centres are listed as operating in the environmental area. They are Waste Management and Pollution Control, Soil and Land Management, Catchment Hydrology, Biological Control of Vertebrate Pest Populations, and Antarctic and Southern Ocean Environment. However, the limitations of the programme also need to be recognized. In particular there are important areas of environmental research not covered by any of the centres so far established. Also, there are many very good researchers outside the centres, and it is essential that a range of funding mechanisms continue to be available to support them. Most substantial improvements in understanding have come from individual researchers pursuing curiosity-driven research. They need the freedom and the resources to get on with it.

Science and Environmental Research: A Feminist Critique

JONI SEAGER

This chapter is the transcript of a conference address given by Dr Seager, published here with only minor editing.

In this chapter I will confine myself to a feminist critique of the role of science in environmental studies, rather than tackle the broader question of environmental research as such.

First of all, it is my belief that the trend is increasingly toward an environmental research agenda that is set by science. We are, I think, moving towards a science-based environmental paradigm. Scientists increasingly have primacy in defining environmental problems, and in constructing environmental solutions. Environmental issues are increasingly assumed to belong in the arena of science, certainly in the view of funding agencies and government policy bureaus. We depend on scientists to tell us when we have a problem, and we depend on scientists to solve the problems they have alerted us to.

In part this is because alternative models seem to be falling by the wayside. Political environmental groups, which in the past often represented an alternative agenda, are moving at lightning pace to recast themselves in a new, professional form. 'Professionalism' in this case is a process which is self-defined by many of these organizations as one which recasts themselves, in the words of the organizers, away from being 'fringe, amateurish, emotional, opposition groups' into streamlined, slick 'expert' organizations. Listening to the leaders of groups such as Friends of the Earth, Greenpeace, and many of the national organizations in the US and UK, it is clear that environmentalism as an opposition movement is undergoing professionalization, and a key feature of this professionalization is increasing reliance on science. Along the entire spectrum of environmentalism, from government policy bureaus to oppositional organizations, the expert structure in environmentalism is tilting towards an environmentalism defined and led by science. This is a trend that I find somewhat worrisome. My concern about the increasing hegemony of science derives from my concern about the nature of science itself, and the terms and the standards that science sets for itself.

So what is the nature of science that I and other critics find troublesome? Science is construed primarily as an exercise in the application of rationality. At its best, pure science strives towards objective discovery and assessment of universal natural truths. Science is supposed to be a rational, truth-seeking, value-free application of the intellect, and, more specifically, science is supposed to be conducted in a realm that is removed from political and social implications and considerations. Science, if you will, is supposed to be conducted in a sterile mental laboratory, and in this laboratory emotion, political considerations and personal values are portrayed as 'pollutants' that undermine the conduct of pure science.

Now, of course, the radical critiques of science that emerged from the political movements of the 1960s and 1970s burst this bubble of complacency about science, and it is now clear to most observers that science in fact is not, and cannot be, value-free or objective or somehow transcendentally neutral. The products of science, the truths that science lays bare, reflect the social or national or racial context of the people who produce the science. Without covering that familiar territory in any more depth, I will take for granted that we agree that science is a socially located enterprise, and that its products reflect the class, gender, race and political profile of the people who create the science.

Despite the advancing critiques of science from the 1970s, gender-conscious accounts were few and far between. The people who were developing the radical critiques of science focused almost exclusively on the relations of science as a product of capitalism, the market relations of science if you will, but did not focus on the relations of science to patriarchy, i.e. science as a gender-specific product. This feminist critique of the gender relations of science has only more recently emerged through the works of people such as Sandra Harding, Nancy Hartsock and Evelyn Fox Keller.[1]

Feminist Critiques of Science

What I will do now is review very briefly the emerging feminist critique of science, and then try to bear specifically on issues pertinent to environmental science. The feminist critique of science starts by asking a basic question: If science is a project of the rational mind, whose rationality are we talking about? The world of science is, first and foremost, a world of men. If you look at the numbers, women's representation in science is dismally low. In the United States (and the figures that I have seen are much the same worldwide) women have the highest representation in the biological sciences, somewhere around 20 per cent, depending on what measures you use, down to about 3 per cent in the hard sciences, particularly physics. But I do not want to dwell on the numbers, because in themselves they are not the problem. Rather, this low representation of women in science is a point which should pique our curiosity. Why is science such a male undertaking? Why have men in science, ever since the emergence of modern Western science, fought so determinedly to keep women out of science? And the bigger question, what are the implications of the fact that science has been a project largely constructed and designed by men?

To start answering these questions I will quote from Evelyn Fox Keller, who is a respected American scientist as well as a feminist critic of science.

> Like most aspiring scientists, I was taught from an early age that thinking scientifically was thinking like a man. I was well trained in the division of emotional and intellectual labour that sets scientific virtues—reason, objectivity, autonomy, power—in direct opposition to the virtues usually relegated to women—feeling, subjectivity, intimacy, interconnectiveness.[2]

Now, feminists who are leery of essentialism, as I am, are quick to point out that it is not an innate characteristic of men to value intellect, objectivity and separateness, and it is not an essential characteristic of women to value emotional expression, subjectivity and interconnectiveness. Indeed there is a great deal of fluidity around gender-related behaviours and characteristics. But without assigning all men the task of carrying burdensome and perhaps unwelcome behavioural baggage, one can none-theless point out that the attributes for success in the scientific world—a privileging of emotional neutrality, a privileging of rationality, of personal distancing—reflect characteristics that do define manliness in our culture. They are characteristics for which men are rewarded and for which women are not. In fact when women do assume these characteristics they are often portrayed as 'unsexed' or 'unfeminine' creatures—we need only think of the stereotypes which persist of the 'frigid female scientist'.

Allowing for flexibility in gender behaviour, it is more accurate to say that the problem is not that science is a product of men, but that it is a product of masculinist culture—a culture which individual men may or may not identify with. I want to make clear that I am not critiquing science as a project of men as men, but rather as an endeavour that is built around a set of assumptions and standards that represent a somewhat distorted masculinism. The association between science and masculinity has a long history. Carolyn Merchant has written the most extensively on the historical links between the emergence of science and male domination of both women and nature. If you look at the founding premises of modern science as she has, it is clear that modern Western science emerged as an enterprise directed at the goal of imposing male order on an unruly, feminized natural world.[3] The writings of Francis Bacon, the 'father of modern science', were predicated on the under-standing that 'Nature with all her children should be bound to our service and made our slave'. Now this is not just a matter of rhetoric from the sixteenth and seventeenth centuries; the notion that science is a triumph of male rationality over female nature is abundantly clear in the halls of science today. Science is still cast in terms of culture versus nature, mind versus body, reason versus emotion, the public versus the private, and in each of these dichotomies the first is associated with men and the second with women. Feminist scientists such as Donna

Haraway have explored the extent to which male metaphors and masculinist presumptions guide the course of science today: for example, the search in physics for the master molecule, or the presumption among biological scientists that in every biological group there will be a dominant male leader.[4]

Implications for Environmental Knowledge

But what about the environment? How do feminist critiques of science help us think about the role of science in the environmental arena? Environmental knowledge is increasingly located in the realm of experts who may have general knowledge, but who lack awareness of local context. Both of these facts—the increasing reliance on experts, and that these experts may be far removed from local context—have gender implications. The increasing reliance on scientific experts in environmental assessment will lead to an increasing marginalization of women; because there are still relatively few women in the ranks of science there will be relatively few women in the ranks of designated environmental experts. So to be accurate about the shift toward a science-led environmentalism, we should say that this is a shift that is privileging not just an outside expert structure, but a *male* expert structure. This fact has set the scene for an environmental drama that plays itself out in a number of guises around the world: 'the reasonable man meets the hysterical housewife'.

The classic model of environmental confrontation in the United States (and speaking with people in other parts of the world, it appears to be so elsewhere) is this. Because women are responsible for maintaining daily life in most parts of the world, it is often women who notice first when things go wrong in the daily environment: the water smells funny, the kids are getting sick more often, the laundry doesn't get clean, and so on and so forth. The women who notice this often start comparing notes. This often leads to the production of hand-drawn maps, let us say of illness incidents in the neighbourhood. Before long, the women start to complain, to 'make trouble'. The experts are brought in, and typically they dismiss the problem out of hand. Once this expert judgement has been pronounced it often takes years, and in some cases several deaths, before a problem is acknowledged, and then only if the experts say it is so. Women who take up an environmental cause, women who notice and complain about environmental problems, are often dismissed by the experts as emotional, hysterical housewives—and that is a direct quote. Grassroots women from around the world all report that the designation of 'hysterical housewife' is a common response to their environmental troublemaking.

Thus we have the stage set for the classic confrontation between men of reason and hysterical women. Now one thing to bear in mind is that it has been 'hysterical housewives' in their various guises who have uncovered some of the greatest environmental disasters of our time, including Love Canal in the United States, Minamata Bay in Japan, and many of the localized toxic-waste sites that dot every industrial landscape. Around the world today it is 'hysterical housewives' who are defending forests in India, who are propelling the green-belt movement in Kenya, and who are in the lead in the fight against military pollution in the United States. Although environmental degradation sometimes does happen dramatically (in the form of a chemical explosion or whatever), more often environmental degradation manifests itself in slow subtle deterioration in the quality of daily life. And despite the fact that women are key environmental players around the world, players with an expertise grounded in community-based knowledge, and despite the fact that we are increasingly recognizing the importance of paying close attention to the local context of environmental degradation, reliance on a cadre of experts who typically have no grounding in local context is moving environmental assessment away from the realm of lived experience. This, I think, is an extremely contradictory and problematic trend.

Science is a decontextualized endeavour, and it is a universalizing endeavour. Scientists strive for the big picture, the natural laws, the universal truth. This is fine except when thinking about the environment, which is, after all, pre-eminently about local context. It is universalizing scientific resource management that has led to the march of eucalyptus trees around the world, with disastrous consequences. It is a universalizing science in aid of empire that has provided the wherewithal to force cash monoculture on tropical soils—an agriculture that can only be artificially sustained by the application of ever more powerful and complex and expensive scientific intervention.

The universalizing proclivity of science also has direct ramifications for women. Women are typically rendered quite invisible by the science of environmental impact assessment when it is combined with the assumption of universal experience that many men take for granted. I will not belabour this point, but it is important to say, and say again, that the impacts of environmental degradation are almost always felt differently by women and by men. This is true whether you are talking about the impacts of extreme environmental degradation, as in the Gulf War or Bophal, or about deforestation in the Himalayas, or about the Aral Sea catastrophe—women and men experience environmental degradation differently. However, I am constantly amazed at and disappointed by the degree to which the truth is not taken into account by the experts who are leading the field of impact assessment.

Conclusions

Finally, while we look to science for solutions to our environmental problems, it serves us well to reflect on the fact that the dominance of the scientific paradigm is what has *caused* so many of our environmental problems. In its recent history, science has not been a politically liberating institution, and the overarching paradigm of Western science is not environmentally propitious. Rather, it is an enterprise based on the domination and control of nature, a comment that has been made by other authors in this book.[5]

Science is an essential component of environmental research and environmental knowledge. But it is only a tool, and a limited one at that. The potential, and present, contributions of science to environmental understanding are constrained by the limitations of Western science itself. The overarching constraint is the recognition that environmental problems are quintessentially social and cultural problems—precisely the realm in which scientific understanding is at its weakest. For this reason alone, we should be leery of assigning science too high a priority on the environmental agenda.

Science can and must play a role in the environmental arena. But given the defining characteristics of scientific enquiry and the nature of the technological application of science, this role can (and should) be only a limited one. The overarching paradigm of Western science—a decontextualized search for objective and universal truths—is not a propitious paradigm for environmental understanding. Environmental knowledge must be, at its core, locally grounded and keenly attuned to the social, political and emotional contexts within which environmental problems arise, are perceived, and are resolved.

The limiting characteristics of the nature of science are in large measure a consequence of the 'culture' of science. When we examine this culture, it is unmistakably clear that Western science is a product of masculinist culture. Attributes for success in the scientific world—emotional neutrality, rationality, personal distancing—are those for which men are rewarded and women are not. Similarly, the application of science and the role of science in environmental affairs is also 'gendered'. In our recent history, and especially in the heyday of global colonialism, science and technology evolved in directions that took control of natural resources away from people who use them for sustenance and survival (usually under the stewardship of women), and put them into the hands of people who use them for profit and vested interests (usually captained by men). Integral to this, Western science has been an enterprise that has too often been used to dislocate local (women's) knowledge of the environment and replace it with exogenous (male) 'expertise'—a process that has caused tremendous environmental degradation around the world. This dynamic continues today. Indeed, the use of science and scientific expertise in environmental impact assessment is becoming particularly contentious, and it may be the single most important factor in widening the schism between women grassroots activists and men in the environmental mainstream. As environmental watchdogs, women see it to their advantage that they have been socialized to listen to their 'gut feelings'; men are socialized to veer away from intuition, and many male environmentalists are reluctant to cast aside the prop of presumed scientific neutrality and expertise.

Men and women in the environmental movement increasingly express polarized views on the appropriate role of science in setting the environmental agenda.

Science is gendered domain. Similarly, the use of science as an environmental tool has gendered implications. In many mainstream environmental and scientific circles, observations such as these are considered to be controversial, and are often scorned and vigorously resisted. This, perhaps, is not surprising. Indeed, the fact that a strictly scientific analysis of environmental issues would lead us away from such a culturally-based observation is both the beginning and the end of the problem.

11

Politics, Science and the Control of Nature

MICHAEL WEBBER

The control of environmental degradation is fundamentally political: conflicts of interest over the distribution of rights and resources must be resolved. Science is also political: it entails internal conflicts of interest and is an element of the broader conflicts of interest within society. How do these two political spheres interact? What are the implications of their interaction for knowledge about environmental degradation in Australia?

A common answer to these two questions is that science is an independent and rational activity. The speed with which information is produced may depend on funding; but the nature of that information is not socially controlled. Science produces knowledge objectively and impartially. Damage to the environment, though, is a management problem. For one reason or another farmers, foresters and miners make resource decisions that are unsustainable. In this model science produces information; if the information is provided to land-users, their decisions are rendered more fitting.

This answer is naive. First, the extent of knowledge about environmental degradation depends upon politicians recognizing that degradation is a problem. If funding agencies—state and federal governments—appreciate that environmental problems are serious, scientific knowledge about them can be accumulated quickly. So what we know depends upon prior land-use decisions: there is a feedback loop in the system. Second, the answer assumes that inappropriate decisions originate in inaccurate or incomplete information. If farmers, foresters and miners knew more, it is said, their decisions would be better. This assumption is convenient for scientists (it justifies their activities) but there is little information with which to support it. Decisions depend not only on information but on values too. Third, the answer assumes that the information produced by science is unproblematic: fixed by the nature of the external world. According to the model, the amount of knowledge produced depends on funding, whereas the nature of knowledge does not. Again, this assumption conveniently supports the conception that science is a rational, objective source of knowledge. But the evidence is that it is not true.

So this chapter seeks to develop a more sophisticated view of the relations between science and policies about environmental degradation. Underlying the chapter is the view that degradation and science are two sides of the same coin; this view contrasts with positive and optimistic attitudes about the potential benefits of science.[1] The first three sections of this chapter introduce environmental science, environmental degradation, and the social context of degradation. The core of the chapter advances three propositions: internal and external limits constrain science to produce information that is inappropriate for controlling degradation; the significance of environmental science to public policy formation is grossly overrated by scientists; and some environmental interventions are tending to make environmental science irrelevant. Throughout the chapter the context is taken to be agricultural land-use decisions as they affect the degradation of land.

A prefatory remark. This chapter comments on scientific thought and the relations between economy and environment. The norms of Western science are norms that I have grown up with and largely accept: they are honourable and I find it hard even to conceive of rational alternatives. Equally I am committed to a materialist, high-resource-using way of life. I am not convinced about the alternatives. So this chapter is not a you-bad, me-good argument. And while I try to understand the limits of science I do not thereby attack the practice of science.

Science and Environmental Science

It is not clear that there is a defensible *a priori* demarcation of the activities that we call science. The distinctions between science and other ways of acquiring knowledge are not obvious. Here I describe what is meant by environmental science and delineate several types of science. Some aspects of the social structure of science as an industry are also described.

Physical environmental science is an activity devoted to understanding how the natural environment works. Environmental science is the activity of understanding the physical and social processes that affect environments. Physical environmental science is about understanding nature; environmental science is about us and nature. The goals do not state what methods are to be used.

Science can be classified in several different ways. A common distinction is between pure science, applied science and technical development. This classification appeals to a 'stages' view of technological change. Another classification refers to the social functions of different activities: production sciences (devoted to improving techniques of production) and impact sciences (that measure the effects of production on environments or societies). Physical and social sciences are also distinguished. These classifications are not merely nominal. A ranking is implied, especially with respect to funding. Young has claimed that the production and the impact sciences attract different levels of funding.[2] Spiegel-Rosing has pointed to the discrepancy between the depressed soft sciences and the richer hard sciences.[3]

However, it is difficult to assess such claims empirically. Funding from the Australian Research Grants Scheme has been dominated by biology, chemistry, mathematics and physics,[4] though to a smaller extent in Australia than in some European countries.[5] In Australia's higher education institutions basic research (equated to advancement of knowledge) accounts for 46 per cent of research funding; the majority of research and development funds is spent on a variety of applied missions.[6] The environmental sciences obtain 13 per cent of basic research funds, but only a very small share of applied research funds is devoted to the environment (2.3 per cent). This evidence of the magnitudes of research funding refers only to higher education institutions.

Funding provides one of society's controls over the rate and direction in which scientific knowledge changes. Society must decide on scientific priorities because of the overall shortage of resources.[7] The evidence about a proposition is therefore only ever the published evidence, and must be tempered by questions about studies that have not been completed or published.[8]

There is a hierarchy of prestige in science, too. Prestige prefers basic to applied science; theoretical to empirical science; general conclusions to specific or local knowledge. In Australia these elements of prestige are related to the issue of international journals: publication in major journals almost necessarily implies abstraction from our particular environmental, historical or social context. (For a wistful description of the preference for the general claims of imported experts to the particular local knowledge of people in northern Nigeria, see Mortimore.)[9]

Nor are these merely classifications of science: they also classify scientists.[10] Citation analysis demonstrates that different branches of science, like science and technology, develop along different trajectories.[11] As a result, much technical change is independent of recent science.[12] Even within disciplines, the basic and applied research communities hardly communicate with each other. The society of scientists is divided.

Science, then, is a social and political activity. It is deeply embedded in our culture: its concepts and methods reflect and are reflected in the concepts and methods of society at large. This is evident not merely in the social analogies drawn from theories of matter,[13] nor in the common mechanistic world view, much less in scientists using competitive models to justify their form of work organisation.[14] More importantly for our purposes, science and society have together adopted a commitment to analysis.

Land Degradation

There is commonly little dispute about the definition of land degradation.[15] Land degradation is damage to the physical, chemical or biological status of land that restricts its productive capacity:[16] a change that makes land less useful to humans.[17] This definition is problematic. First, soil erosion, mass movement and solution are normal processes of landscape evolution.[18] Australia's soils and groundwaters are naturally saline: they contain geologic salt, and high rates of salting are encouraged by rapid evaporation.[19] By definition such natural processes constitute land degradation if they reduce productive capacity. Second, it is not clear whether 'capacity' refers to actual or potential productivity. If capacity means actual productivity, unused land cannot be degraded. If potential productivity is the test, the conditions that constrain potential must be specified: potential under what circumstances? Third, land degradation depends on technology: when we invent a cheap way of fertilizing land, farming systems that mine nutrients may no longer degrade.

Land degradation has several associations. It may refer to changes in the quality of land (under standard weather conditions); to unsustainable changes in land quality (yields are tending to fall under standard weather conditions); or to suboptimal changes in land quality (that do not maximize the present value of future expected incomes). All three definitions surface in the literature. What is degradation by one definition may not be degradation by another. Thus 'lower' land quality and improved technology may offer the same yield as 'better' land quality and simpler technology. As we improve techniques to obtain higher yields from given quality land, sustainable changes in land quality may involve 'deteriorations' in land quality.

Changing Environments

Land degradation can be understood at three levels. At the technical level, degradation is a matter of the physical environmental processes that remove soil or vegetation or add undesirable properties to soil. At another level, land degradation can be understood in terms of the decisions that people make about managing their land. Physical processes cause land degradation in the context of management decisions. This is the social level at which we analyse farmers and their socio-economic circumstances. At the third level, land degradation is one of a suite of environmental changes: it exemplifies a disregard for nature that originates deep in the characteristics of our society. Farmers' decisions and social circumstances reflect these deeper characteristics. In this section, knowledge of land degradation at each level is briefly reviewed.

The first level is the technical level. It is thought that we understand the main physical processes of land degradation fairly well, and have some handle on its extent.[20] The main sources of land degradation in Australia are salinization and erosion by water and wind.[21] Water and wind erosion require treatment on 796 thousand square kilometres of non-arid Australia (44 per cent of the land in use) and on 1850 thousand square kilometres of arid Australia (55 per cent of the land in use).[22] Salinization of soil and water has occurred as water tables have risen because of increasing rates of groundwater recharge and as topsoil has been eroded to expose saline subsoils.[23] Perhaps 42 thousand square kilometres are affected by dryland salinity and 15 thousand square kilometres by wetland salinity.[24] However, Burch *et al.* claim that we have 'considerable scientific understanding of land degradation processes now at our disposal, though many gaps remain to be filled . . . Scientific know-how is not necessarily the rate-limiting step in preventing land degradation. Restrictions exist in the difficult steps needed to employ the knowledge'.[25]

We have less information about the effect of current agricultural practices on land degradation. Three periods of white occupation of arid lands were typified by: rapid growth of stock numbers; collapse of numbers in the face of low prices and environmental deterioration; and stabilization of numbers.[26] Western New South Wales was severely degraded by the early twentieth century; yet by the mid-1970s recovery was obvious.[27] Perhaps European agricultural practices are the main source of degradation:[28] Galloway estimates that rates of erosion are now sixteen times their pre-European rate.[29] Or perhaps changes in annual rainfall have played a role too.[30] We may know the processes but we have less information about their relative sig-

significance—particularly as it shifts over time and varies over space.

The second level concerns management practices and their social controls. It seems reasonable to suppose that farmers make rational decisions, given their resource constraints and uncertainty about prices and weather. However, there is no reason why rational decisions should foster sustained yields, much less constant land quality. If the discount rate is positive, future income is valued less than present income; so under constant prices an optimal land management strategy invests less in land quality than is needed to maintain yields. Furthermore, farmers who value present income much more than future income and farmers to whom the cost of capital is high degrade their land faster than others. No matter what the quality of science, farmers may quite rationally degrade land: information need not be the constraint. So land degradation as a physical process is different from land degradation as a social problem;[31] the problem occurs when farmers degrade their land at a rate that exceeds the social optimum (defined when information is perfect and individual discount rates equal the social rate). However disquieting such an approach may be, it does identify the rationality of land degradation. Only when the planning horizon is long and the discount rate low is land conservation a paying proposition to private farmers.[32]

Though the rational expectations model defines land degradation and recognizes its rationality, it is an unsatisfactory guide to social practice. It avoids issues of pricing.[33] It ignores the offsite costs of land degradation, which are larger than the productivity losses caused by soil erosion.[34] The price of an old tree is not counted. It assumes that land-users own the rights to environmental quality.[35] And in using discounted future values to measure the cost of degradation, it simplifies the riddle of intergenerational equity, assumes that there is a perfect forward market for land quality, and rationalizes present greed.[36] More importantly for our purposes, the model neglects the social and economic influences on farmers' behaviour: water prices, settlement schemes, international prices, fertilizer bounties.

The third level asks what kind of society can contemplate an optimum rate of land degradation. Two characteristics of

Western European, North American and, increasingly, global cultures are crucial.

The first characteristic is a pair of drives: to understand what and where we are, and to master nature. Dorn recognizes both drives when he distinguishes a Greek tradition of un-funded, private, useless science that seeks understanding from a hydraulic tradition of bureaucratic, applied science devoted to controlling the forces of production in civilizations like ancient Egypt, Maya, Sri Lanka and China.[37] In Christian tradition, we have an absolute mandate to subdue nature, which was created for us: we cannot sin against the natural world.[38] Both drives survive in the attitudes of scientists. But whatever their source, the effect is the same: try to control nature. 'Man against nature. That's what life's all about.'[39] The attempt underpinned the European response to Australian difference: change the land to make it like Europe.[40]

The second characteristic is that our societies are organized by capitalist production. The theoretical basis for this characterization is disputed; all I draw from it is the claim that production is organized to accumulate capital. The entire basis of the economic system is growth. Young writes that our societies are hooked on growth, since more consumption is the only benefit provided to most people.[41] My claim is stronger: growth is not merely a drug to which we are addicted; it is the whole basis of our economic organization.

At the third level, land degradation is assessed in terms of these two characteristics. Are capitalist organization and the drive to understand and master nature necessary and sufficient for land degradation? Or does degradation derive from less central characteristics of our culture and society? Indeed what are the relations between capitalist organization and the drives to understand and master? Certainly the mastery of nature provides a means of capital accumulation, while profits and growth provide questions for science to answer, as well as money and techniques to seek answers. But whether different forms of knowledge are compatible with capitalism or whether modern science is compatible with different forms of social organization is not clear. A separate question concerns the necessity for deep ecology.[42] Or can conservative land ethics be derived from anthropocentric approaches to the environment?[43] Given these uncertainties, Figure 11.1 describes the central relationships that may be implicated in changes in land quality. This figure provides the background to the comments that follow about science and land degradation.

Constraints on Environmental Science

First, I claim that science produces knowledge that is inappropriate for solving the problems caused by land degradation.

Scientific knowledge of land degradation is not perfect. However, it is thought that we have good theory about processes and a large-scale appreciation of the extent: science is in not too bad a shape. What is lacking is an understanding of the interventions and of the social science of land degradation. Research on conservation lags that on erosion: there has been little effort to model the effects of soil conservation measures on erosion rates.[44]

The social processes governing land degradation are poorly understood. Physical scientific approaches to land degradation emphasize biophysical rather than socio-economic causes. They blame ignorance, laziness, irrationality, lack of environmental awareness.[45] The following comments by researchers in the field are typical:

- Three factors limit prevention of land degradation: insufficient monitoring of processes and extent; inadequate application of information to control degradation; inadequate appreciation of costs.[46]
- Some landowners are prudent and realize the value of conservation measures; others seek the fast buck.[47]
- The global problems of desertification require research to classify and record grazing lands; identify early deterioration; analyse ecosystem structure; understand the dynamics of plant populations and communities; improve management decisions.[48]
- The steps in a water quality management programme are: learn the system; list the beneficial uses that are to be protected; choose indicators of protection; choose strategies to protect; monitor the effects.[49]

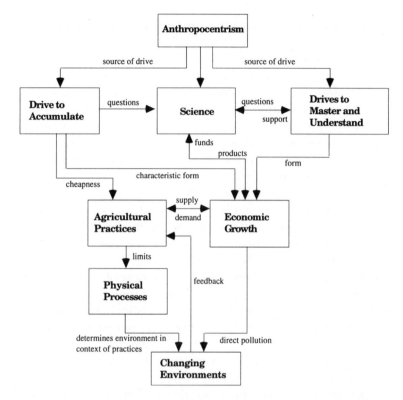

Figure 11.1 Relationships between Western culture and changing land quality

- Measures to control salinity are: control of recharge; control of discharge; water treatment.[50]
- Research on soil erosion seeks suitable conservation measures. Validate soil loss tolerances. Design strategies for erosion control. Design conservation measures to achieve these strategies by manipulating equations for critical slope length, channel flow and wind flow.[51]
- Control of environmental degradation requires land-use controls; integrated management of land and water resources; recognition of the socio-economic problems that are caused by poor practices.[52]
- Williams recommends a series of research projects about salinization and waterlogging in the Murray–Darling Basin; he includes only one social proposal—on the legal and social implications of salinity.[53]

- Salinity can be controlled in dryland areas by tree planting and use of deeper-rooted plants. Water conservation, better drainage and improved levelling of land are needed in irrigation areas. The impediments to salinity control are lack of awareness; lack of understanding; financial constraints; inequities when cause and effect are separated in time and space. Practical requirements include a Salinity Control Board; interagency co-operation within government; an education programme; research and monitoring; loans for incentives and land purchase schemes.[54]

Three points need to be made. First, for the most part the writers display an abysmal lack of understanding of the social and economic constraints under which farmers make decisions. The proposals are technical, and ignore the institutional constraints on land users. Second, they implicitly accept that the rate-limiting factor in preventing land degradation is information: 'Information is the key to efficient resource use.'[55] This was a theme of early environmentalist classics,[56] but this paradigm has been replaced: the environment is a political as well as a technical problem. Third, the technical solutions have been derived from a physical understanding of systems, rather than in conjunction with land-users.

Quite apart from the rationality of land degradation to farmers, many features of Australian society also encourage cavalier attitudes towards the land. Historically the Australian focus on land development has encouraged farmers in good seasons to extend agriculture into inappropriate lands. Private land ownership and the failure to protect common-law environmental standards have invited land-users to regard environmental quality as theirs rather than communal. High interest rates discourage long-term land improvements. With relatively infertile soils and marginal climate, Australia's comparative advantage in agricultural commodities has resided in high capital and land inputs per farmer and in low land prices, precisely the conditions that encourage the mining of soils. In any event an industry that competes on the basis of price is driven to minimize production costs rather than to maximize the qualities of inputs (including land); farmers are compelled to reduce their costs of production to match those of the most degrading practices overseas. Little is known about the relative significance of these factors. Still less is known about manipulating market conditions to achieve desirable outcomes for land quality.

We have a scientific understanding of what is going on and its rate. We do not understand how to apply the technologies for land conservation. Nor do we understand the social context of land degradation. The limits to land conservation are not science, but technology and social science. This state of affairs is consistent with the cultural characteristics of Western science:

1. Hierarchies of prestige, funding and believability mean that knowledge about physical processes can be accumulated more easily than knowledge about social processes. Knowledge about increasing production is more easily obtained than knowledge about conservation.
2. A preference for basic research means that applied, locally-specific research is hard to defend in terms of scientific respectability. It may hardly be publishable and has messy outcomes: clean, interpretable experiments are not possible in this work.
3. Patterns of funding and research reflect a cultural predisposition to tinker with technology rather than to alter the social conditions that promote a problem: fund trees and contour ploughing rather than alter the social conditions that foster land degradation. I said cultural predisposition; yet these are issues of power and politics. We tinker with techniques rather than address private property and degradation, the rights to environmental quality, the pressures on farmers to degrade their land.

Environmental Science: Problematizing Land Degradation

We do not have good estimates of the costs of land degradation.[57] The cost of repairing land degradation in Australia may be $2000 million (present dollars), just under 10 per cent of the annual value of agricultural production.[58] Over two hundred years that works out at $10 million damage per year. This land degradation may cost Australia $600 million a year in lost production.[59] By comparison, crop productivity losses in the USA due to soil erosion have been estimated at $40 million per

year (nearly $70 million in present Australian dollars).[60] Either
Australia is much worse off or the data have a radically different
basis! It is estimated in the USA that the annual off-farm costs
of sediment and pollution entering streams from agricultural
land were fifty times the productivity loss. If we reduce this ratio
by an order of magnitude to be conservative, the annual on-
and off-farm costs of land degradation to Australia are about
$3000 million, some 15 per cent of farm production. These
calculations indicate that land degradation is a problem even
economists might worry about.

So is land degradation a problem, and in what senses? Conventional opinion observes that land degradation is extensive,
but many economists argue that it may not be an important
problem.[61] They argue that land quality is changing; perhaps
land quality changes are not sustainable; but there is no
evidence that rates of land degradation are not socially optimal.
(Data on off-farm costs of land degradation raise the estimates
of costs of land degradation substantially; but such costs have
not entered the economic argument.) So why are governments
intervening? And why do scientists support intervention and
regard land degradation as a problem? What has this public
policy got to do with science?

First, the scientific push to intervene. Many scientists are
concerned at the physical changes taking place in Australia's
environments. The indignation in Chartres, Galloway, Neville,
Robertson, Smith, Smith and Finlayson, and Williams is clear,
if not explicit.[62] This indignation is independent of economics
and is consistent with widespread public indignation about
global environmental problems. But scientists also have social
reasons to encourage intervention. Techniques of controlling
degradation are a fruit of environmental research, a public
benefit to which scientists can point to justify their work and
their public funding. As fluoride was the public health benefit
of dentistry,[63] so land degradation elevates those who work on
salinization and soil erosion. The effort is not purely 'scientific';
how could scientists justify their recommendations in the face
of the comments of economists?

Why then do we have public policies to offset land degradation? In part, politicians respond to scientific pressure. They
are sensitive to environmental concerns: remedies for land

degradation are pretty cheap in comparison with reducing urban pollution. And no doubt there are residues of a conservative ethic of land management among farmers and some politicians. Most significantly, some of the common preventative measures—stubble retention, whole-farm planning, trees in recharge areas—are justified outside soil conservation.

Consider the example of water management in Victoria's irrigation areas. The control of salinization by reducing groundwater recharge rates is touted as a reason for increasing the price of water and encouraging laser levelling of land.[64] Yet changes in water pricing policies have other justifications. The former system of charging for irrigation water encouraged farmers to use all the water that was supplied to their district.[65] In the early 1980s the Rural Water Commission and others began to question this policy. The system encouraged inefficient use of water. A new view developed: water should be priced to encourage efficient resource use. Farmers in new schemes should be charged the full cost of supplying water, while those in existing schemes should be charged the variable cost.[66] Prices should be raised to see what the market would bear, to find higher value uses for water and to raise the degree of cost recovery.[67] Given the budgetary constraints of state governments in the 1980s and 1990s and a political climate in which users pay for public services, it is hardly surprising that the Water Act 1989 incorporates the recommended changes in water pricing for irrigation areas. The fact that higher charges for water may promote water conservation and diminish groundwater recharge seems in the discussions of the early 1980s to be a fortunate by-product rather than a reason for the policy: it is also a by-product that has received a lot of attention after the event.[68] Conservation was less a matter of public policy than an effect of other policies.

So the contribution of environmental science to public policy is limited. Scientists have interests other than simply science; and governments respond to pressures other than simply the facts of land degradation.

In any event, the capacities of science are unsuited to the tasks of managing conservation. Remedies are vague; data are uncertain; scientists find it difficult to modify their careful science to acknowledge noise in data and the normative demands of management; the relations between physical science and the contexts of policy development are misunderstood by scientists; the strategy of analysing manageable subsystems is not suited to providing conclusions about prespecified whole systems; the problem of modifying qualified, quantified research findings by experience, common sense and intuition: all these problems have plagued even the most careful environmental management programmes.[69] Scientists have proved incapable of providing professional advice to a client who evaluates that advice commercially.[70] There are special problems for scientists whose conclusions conflict with the wisdom of their funding agencies.[71]

Science is unlikely to figure large in any public debate over remedies for land degradation. (Debate over medical studies did not count in arguments about fluoride.[72] Equally, science has been quite unimportant in campaigns about cattle in the Victorian Alps.)[73] If physical scientists and environmentalists try to persuade governments to require conservation programmes that will cost farmers money, such a debate will arise. However, the debate is likely to be about economics, ethics, property rights; the 'facts' of land degradation are unlikely to be in the forefront. The science of land degradation is technical, and most people are not educated to understand it. Even extensive argument is unlikely to include much scientific input.

Environmental Science in Environment

Our attempt to control nature depends on its predictability: given the short-run fluctuations of weather, nature changes slowly in relation to the spans of human lifetimes. That is why induction works: it is not logically defensible, but it is empirically reasonable. Something that happens twice is probably true.

We have made two dramatic interventions that alter this assumption and make environmental science increasingly difficult. We are altering global climates. The greenhouse effect and holes in the ozone layer alter global climates at rates that are orders of magnitude faster than natural changes. Rapid changes in land quality are in turn being induced by climatic changes. And we have begun to introduce our own species:

biotechnology means that we are creating forms of life. On both counts the world is becoming our creation: a randomized Disney World with fractals. Outside no longer differs from inside: the meaning of nature has changed.[74] Of course, the changes we have made are not entirely predictable; so we shall certainly need to continue to study the world.

All environmental processes are now the subject (or potentially the subject) of human intervention. Whatever happens, we might have caused it. Once the human contribution has become more than trifling, the evolution of the new landscape is no longer natural, and catastrophes lose their status as simple acts of God. If there are floods or droughts, hurricanes or calms, hot weather or cold: all might be our effects. Science, having claimed to be the basis for our capacity to control nature and unleash technology for human benefit, will have to wear the blame.

The basis for study of the world has to change. The warranty that on average the environment will remain more or less as it always was has expired. The basis for induction is lost. People have begun to understand this. The lower Mississippi River, for instance, 'is an extremely complicated river system altered by works of man. A fifty-year prediction is not reliable. The data have lost their pristine character. It's a mixture of hydrologic events and human events ... This is planned chaos.'[75] McKibben has reacted strongly to this effect.[76]

Whatever we study out there will not be nature—our environment. The study of soil erosion and of land degradation will be no different from studying runoff in a shopping centre. Necessary perhaps, but not the stuff out of which ambitions to understand the universe are built. Environmental science becomes impossible: there will be no environment to science.

Conclusion

Environmental science as it is now organized has not been capable of explaining the social bases of land degradation; it has not figured large in developing an agenda for public policy about land degradation; and as things are going, it is running out of an environment to understand. In one sense these conclusions counter scientists' pretensions to contribute on a grand scale to land conservation. But they should not be regarded as claims that science is irrelevant or that tinkering does not have a part to play in preventing environmental damage. Rather I would argue that scientists could play a role in conserving our land. One vital task is to use the physical insight that the non-renewable resource of soil is being destroyed in order to undermine the concept of an optimal rate of land degradation. Another vital task is to disentangle the forces that underlie land degradation: will tinkering resolve the issue or do we need to be more fundamental? My pessimism is that these are not the sorts of jobs that science is constructed to do.

Part III Action: Dealing with Land Degradation

H ow might a different approach to knowledge and an ethic based on values appropriate to our long-term needs, as discussed in previous chapters, influence the way we behave as individuals and respond as a community to our environmental problems? To examine this question concretely and specifically, we turn in this final section to land degradation and sustainable agriculture.

As noted in the introduction, the picture painted in the major national survey, *Land Degradation in Australia*, in 1984 was grim. The case study in this book examines one aspect of land degradation only, that related to farming land. Here, however, the picture is no less alarming. *Land Degradation in Australia* found that 51 per cent of all agricultural and pastoral land required treatment, and more than half of that land, amounting to 29 per cent of all agricultural and pastoral land, required treatment with conservation works in addition to improved management practices. In assessing the urgency of this work, the report identified that management treatment was most urgently needed in the Darling Downs in Queensland and the Mallee and mountain areas of Victoria, and that treatment by works as well as management was most urgently needed in the Hunter Valley in the central tablelands of New South Wales, in northern Victoria (for salt-affected lands) and in arid areas in western New South Wales and the Gascoyne Catchment in Western Australia. Land-use change was most urgently required in the Alice Springs district and in the southern Kimberleys.[1] Some of that action was identified as being essential in the ten years from 1975 to 1985 and the rest over the twenty years 1975 to 1995. We are now very close to 1995.

It is likely that there has been further deterioration since 1984. There have, however, been some serious attempts to deal with specific aspects of the problem. Four types of measures will be discussed in this section: salinity programmes, Landcare and native vegetation retention as government programmes; and conservation farming as an industry programme.

The questions that this section will explore are: Have these measures been effective? If so why, and if not why not? In what way has knowledge influenced behavioural change? How adequate is the knowledge base? Are the values informing the way we generate knowledge wrongly influencing its direction and dissemination? Are the values and belief systems of farmers and others a barrier to change? If so, how can they be modified? Are the structures within which research, government programmes, market forces and farming practices operate a major impediment to the development of sustainable agriculture in Australia? If so, how can they be changed and how might a different theory of knowledge and a new or emerging environmental ethic assist that change?

Ecological sustainability is the focus of the enquiry. There is perhaps no better arena than agriculture and farm lands in which to pursue it. Food and fibre production is an immediately recognizable need for any society; we do not have to debate the necessity of an agricultural industry. The need to deal with land degradation and to develop a fully sustainable agricultural practice is equally evident, a long-term environmental goal agreed by all those involved, from farmer to conservationist. There are, however, two quite different perspectives on the problem: that of the farmers and agriculturalists on the one hand, and that of environmental researchers and policy-makers on the other.

The chapters that follow look at the problem from these two different perspectives. Adrian Egan and David Connor begin by placing agricultural productivity and sustainability into a broad contextual framework, properly insisting that agriculturalists cannot shirk the responsibility for food and clothing production, and illustrating the many dimensions of agricultural demand and supply and the difficulties of defining agricultural sustainability. Leigh Walters and Albert Rovira emphasize the difficulties of the farming task and, using conservation farming as an example, show how such new approaches require more skills of farmers, many of whom are ill-equipped to respond properly. Returning to the questions of knowledge and research discussed in Part II, they illustrate why linear models of the transfer of research findings do not work, and why a different model for sharing knowledge is needed in which each participant—farmer, extension officer and researcher—has an equal role. Peter Small, speaking directly on behalf of farmers, makes a personal plea for better understanding of the problems of farmers and for a community effort to bring the perspectives and energy of town and country together to deal with them.

John Bradsen introduces a different perspective, arguing that there is little evidence to support the view that we can achieve the changes we require if we rely on the initiatives of farmers and agriculturalists alone, however much these initiatives should be encouraged. He points out that we should be clear about our goals, arguing that it is the protection of biodiversity that really matters. He looks at the means available to us to change behaviour, examining in turn science, ethics, economics and the law, and gives examples of South Australian programmes backed up by intelligent legislation as one way forward. He also introduces Landcare as a promising but flawed model for bringing farmers and others together to solve collective problems. Landcare and salinity programmes become a major theme around which the concluding chapters focus. Evan Walker demonstrates the effectiveness of a carefully planned programme—as was the Victorian Government's salinity programme—which combined strong collective government commitment and local responsibility firmly vested in those most directly affected. He also shows how the salinity programme became the model for the Victorian Landcare programme, and how this in turn became the model for the Australia-wide Decade of Landcare.

In the last chapter of this section Sharman Stone brings the two viewpoints together in an essay remarkable for its marriage of the insider's and outsider's views (the direct experience of the farmer and the insights of the analyst and sociologist). She also returns to questions of attitudes and values to show how the long-held credos and sustaining myths of farmers, however legitimate and understandable their origins, are now becoming serious barriers to the development of more sustainable practices. She contrasts the great success of the salinity programme with the limited success of the Landcare programme to point out the conditions necessary for farmers to come together and work with others to solve catchment-wide problems.

12

Productivity and Sustainability of Agricultural Land

ADRIAN EGAN AND DAVID CONNOR

Enquiries into land use and the practices involved in agriculture have brought forward some sixty major reports to Australian governments over the period since World War II. The Western Australian Select Committee into Land Conservation is among the most recent to record both failure of existing processes to end land degradation and success in the use of existing knowledge in specific areas to accomplish rehabilitation. The failures are not caused by lack of improvement in recognition of the forces at work or of understanding their more obvious consequences. They arise, rather, out of confused objectives, short-term priorities, lack of access to tested remedial measures, and limited availability of procedures for monitoring change. These latter are essential if land managers are to recognize early indicators of success or failure in response to the measures being applied.

This chapter deals with the realities faced in the management of agricultural land to achieve the twin objectives of productivity and sustainability. (Productivity expresses output per unit of input, while sustainability is the capacity of the system to maintain productivity over time.) To limit the discussion, the chapter concentrates on food production systems. Whether the approach is anthropocentric or based on geomancy, as is P'ungsui,[1] the method of influence has usually been the same—a limited number of people who 'know' provide the guidelines for others to follow. The task is to understand the demands on land, develop realistic goals, and place the tools of measurement and the basis for judgment in the hands of skilled land managers.

Our problems are the practical ones of finding suitable and diverse approaches that enable land-users to meet continuing, real needs in the face of real costs. The challenge is great, and must be met within a variable environment that is incompletely understood, and in the face of changing, usually increasing, demands of society. 'Perceptions' and 'holistic' approaches are not equal to that task. As knowledge will always remain incomplete, we must do the best we can: exploring systems, ensuring good communication and maintaining research.

The Present State of Land Use

There are those who contend that we practise a European agriculture in Australia and that it was the 'wrong' choice right from the time of settlement. The reality is that we have always practised an adapted agriculture that evolved, by innovation, out of successive lessons and demands in the search for greater and more reliable production. Each step was set in place by individuals as their best bet, given the incomplete knowledge of the time. Those individuals tried what they had seen elsewhere, and also devised potential solutions from what was then understood about the component processes. Inevitably, mistakes were made and sub-optimal directions were explored. There is, however, no going back to undo what in hindsight may seem misdirected steps in the evolution of Australian agriculture.

A major question here is what features of agricultural systems promote productivity and sustainability and how current methods compare with those they have replaced. It is important to emphasize that modern agriculture cannot be identified as a single system involving uniform application of knowledge, machines, and agrichemicals. Diversity of farming practices occurs because soils, climates, access to markets and technology, and cultural influences differ greatly between regions. Farmers are numerous and independent operators who seek the most appropriate practices for their particular combinations of land and other resources. Similar diversity exists within the research and extension professions. This diversity has been a priming force for continuous streams of innovation and change.

Advances in scientific understanding, in particular of the chemistry and biology of soils, of the nutrition, metabolism, and reproduction of plants and animals, and of life cycles and vulnerability of pests and diseases, have allowed significant refinements of many long-standing practices and the development of new approaches. In contrast to earlier times, agricultural practice has become strongly chemically based, not just in the agrichemicals that it uses, but also in the entire approach to the understanding and management of the biochemical processes that comprise productivity. Understanding of climate and weather has also advanced significantly. Measurement of the status of major climate systems such as ENSO (El Niño southern oscillation) provides the basis for long-term seasonal forecasts for our region that now assist in setting broad production priorities. At the same time, however, the forecasting of weather, the day-to-day variation in meteorological conditions that dominates the management of agricultural systems, remains an elusive goal.

Ecosystems are dynamic, and continue to change regardless of any human perception, even over several lifetimes. Land degradation is often the result of a failure to examine whole farming systems in the context of the hydrological and nutrient cycling systems in which they operate. Ecosystem function can be described by identifying pools of those physical and biological components and the rate constants for each of the transfers of materials and energy between them. Input and output rates are recognized externally and have received most attention in the past because they define the performance and efficiency of the system as it responds to both minor cyclic changes in environment and to extreme variations of climate or management. Although manipulation by people may be directed to a single component pool or process, the effect of management is felt generally throughout the system. A thorough understanding of the system demands the appreciation of internal transfers and intermediate states, because they determine the stability of the system and how it responds to change. That understanding can also provide tools for monitoring the 'health' of the system, a theme we will return to later.

When a new state has been established by intervention for greater production, it does not follow that the system is less sustainable. The maintenance of productivity depends on the ability of the manager to monitor the consequential changes and adjust inputs that maintain the new pool sizes and rate constants. The many changing demands placed on land and water resources require not the maintenance of the status quo, but the retention of the resource base to allow further adjustment to inputs to achieve a new, desired output.

Productivity and Food Production

In 1798 the English economist Thomas Malthus analysed the links between population size and food production. At that time, Ireland was surging towards the potato famine with a population growth rate of 1.2 per cent per year, while European and world populations were increasing at 0.3 per cent. Malthus concluded that the potential for food production was limited and, if human populations were not controlled by 'prudence', they would inevitably be controlled by starvation, war, or pestilence. He was not optimistic about preventive restraints, and he and many since have been led to express dire views of the fate of humankind.

Malthus was not able to anticipate the extent and nature of technological advances that industrialization would provide, and he misjudged the potential carrying capacity of Earth. He also misjudged the potential for avoiding pestilence and disease and for prolonging life expectancy. We can follow the changes in history books. The second half of the twentieth century has seen a massive increase in world population and attendant food requirement that is unevenly distributed over continents, populations and agricultural systems.

Prior to 1940, when the world population was near 2.2 billion, farming was conducted throughout the world with 'organic methods', and grain yields were only slightly greater than during medieval times. Since World War II production has increased through expansion of the area under cultivation, improvements in cultivars and management practices, including irrigation, pest and weed control, mechanization, and, most important, through improved plant nutrition. World grain production increased almost threefold from 640 million tonnes per year in 1934–38 to 1740 million tonnes in 1988.

The present technological level of farming systems in developed countries is variously characterized as 'conventional' or 'modern'. Developments important in modern agriculture are: increased scientific and technological knowledge; replacement of human and animal power, principally by internal combustion engines; and widespread use of fertilizers and biocides that, respectively, promote productivity and protect it. Replacement of human labour and animal power by machines changed the scale, speed, and timeliness of operations, as well as providing sharp increases in productivity and releasing large areas of land from production of fodder for work animals. The important benefits of modern agriculture are enormous increases in productivity and in the stability of food supplies.

Despite a doubled population and the decline in the area of cultivated land, developed nations are now well supplied with food. Some, including the USA and members of the European Community, are plagued with surplus production. Other countries such as Australia, with its small population and large land area, look to the export of agricultural products to balance the imports of manufactured goods. Because agricultural yields have been brought closer to limits imposed by climate and biology, the potential for further increases in food production in these countries is now smaller than before. In most cases, however, actual yields are still below those attainable with more intensive methods, indicating that significant reserve capacity remains for food production.

Developed countries have encountered difficulties in balancing supply and demand for agricultural commodities during the period of intensification since 1950 because rapid increases in production outstripped demand. In the 1950s surpluses of traditional exporters were absorbed by developing countries, but that market then diminished with the Green Revolution. In the 1970s demand surged again due to population growth, higher oil prices, and droughts in Asia. That brief period of high prices for food stimulated fertilizer use around the world with the result that exporters again faced declining markets during the 1980s as Europe, India, China, and other importers became self-sufficient and, in some cases, even turned to exporting.

Supplies of farm commodities in excess of demand have caused real prices for agricultural produce to lag behind the returns achieved in other economic sectors. This continues a trend that began in the 1870s when the produce of Argentina, Australia, and North America came on to world markets. Where farmers in developed nations such as Britain, Canada, the USA and Australia have maintained their incomes, in spite of declining prices, it has been through dramatic increases in labour productivity achieved with larger machines and larger farms. Others, in Japan and France, for example, have been protected with subsidies and by restrictions against cheaper imports.

Sustainability

The nature of agricultural businesses

It is important to return always to the fact that agricultural land is managed in discrete units by individual owners or operators. While there is commitment to profitability and sustainability, there is no uniformity in the balance. The balance is determined separately in each case by the financial status and requirements of the individual operators.

More choices of technology and practices are needed to meet the combined objectives of improving productivity and the marketability of products. In seeking these, it is essential that future developments in production and processing meet the full costs of the processes. Delivering undisclosed costs on neighbours, on the 'commons', and on the future, is poor economics and socially unacceptable. Outputs from one system are usually inputs to others, where they frequently reduce productivity. Some undesirable outputs arise on the farm from practices introduced to increase agricultural productivity. Others arise during processing beyond the farm gate. Waste disposal and pollution control must be included in industry balance sheets and environmental impact analyses. These costs have not been identified well in the past. There has been a tendency to label them as 'externalities' and then to ignore them.

Producers must respond to market signals. On the one hand, there is discrimination in some markets against products from systems regarded as having undesirable environmental, animal-handling or food-contamination aspects. These pressures may be deserved, but in some cases discrimination has been exercised without any sound basis and must be countered, at expense to producers, by education of the consumers. The other side of the issue, of course, is that some consumers will pay premium prices for preferred products or methods of production. In terms of national markets, that advantage generally flows to some localized producers. However, in international trade the same advantage can flow to individual countries. Australia, with its dominantly extensive, low-input agriculture, is in a good position to offer such preferred agricultural products at premium prices to other parts of the world.

Environmental impacts

Across the spectrum of food production systems, there are concerns for the future of agriculture because production is either insufficient or is perceived to be unsustainable. In developing countries the concern is for nutrients and other inputs, and for new lands to develop for agriculture in order to meet increasing demands for food and export income. In Australia, the expansion of agriculture into more fragile environments demands close attention.

In developed countries, particularly those with exportable surpluses of agricultural commodities, attention is turning to the cost of 'excess' production to the environment. Efforts are being made to retain as much land as possible in nature reserves so that genetic resources can be protected and the land can be enjoyed for its other values. In many cases the land most likely to support highly productive agriculture is kept in production, so the sustainability of those systems becomes a critical issue.

Increased productivity has not been achieved without social or ecological costs. Change was essential for increased levels and efficiencies of food production, and in addition some disruptions occurred through poor choices and mistakes, oversights, or the inability to account for by-products of the systems. The range of concerns of both the public and of agriculturalists is wide and largely justified. They include:

1. pollution of surface and subsurface waters by agrichemicals, their residues and contaminants, rendering them unsafe or unsuitable for other uses, and disturbing natural systems;
2. contamination of products with residues of agrichemicals, rendering them unsafe or uncertain for consumption;
3. the necessity for farms to become larger and less labour-intensive in order to produce at competitive prices;
4. making agrichemicals a significant part of the cost of production and thereby an economic strain in seasons of low yield or low returns and a drain on energy resources;
5. providing inputs that allow farming to extend to fragile land, increasing the competition with nature conservation and, when poorly maintained, causing land degradation;
6. salinization through failure to provide adequate drainage.

These concerns emphasize the need for greater understanding of the operation of agricultural systems and their interaction with other land-use systems and society generally. In this way, management has the opportunity to overcome the problems and society can understand the costs involved. One outcome of this analysis is the identification of indicators of change to provide signals for effective management of complex systems.

Indicators of change

Productivity is an important indicator of change that relates directly to economic performance and the needs of society. However, it may change too slowly or be too much obscured by weather-induced variability in output to provide an effective basis for system management. We need to establish more effective short-term indicators. We can measure change in soil pH, or in cation exchange capacity, or in structural integrity of soil aggregates, or in the spectrum of soil flora and fauna; but how much change in these factors is indicative of incipient degradation or, alternatively, of success in land rehabilitation? How far can we go in establishing a confidence to transfer what is learned by study in one place to another on the basis of some simple key measurements? How much of this can we establish from remote (satellite) sensing as opposed to field sampling and subsequent laboratory analysis?

Conservation tillage provides a useful example. Under this system of management, tillage is largely replaced by the use of herbicides so that in comparison to conventional tillage significant changes develop in surface soil properties. These have been identified by Uren as: increase in bulk density or decrease in total porosity, increase in soil organic matter, increase in soil strength, increase in macroporosity, increase in aggregate stability, increase in infiltration rate, lower soil temperature, increase in soil acidity, and increase in soil fauna.[2]

Thus conservation tillage, at the expense of increased use of herbicides, makes land less prone to surface erosion, increases water-holding capacity, and slows down percolation of water through the profile, which may reduce accession to the water table. In the long term, reduced erosion will assist productivity and likewise, along with lessened drainage, will reduce off-site effects currently associated with conventional tillage practice. However, in the short term there is no consistent effect on yield, so for individual producers there is little incentive for change. The environmental problem is that once decline of yield is recognized, recovery of the system may be unaffordable or impossible.

The Continuing Challenge

Change will continue, but the nature, timing and rates are unknown or are poorly predictable. Not all change will be unfavourable, and we may be able to harness some to the advantage of greater productivity and sustainability. Agriculture must be both imaginative and realistic in identifying opportunities and developing long- and short-term strategies to meet the needs of the growing population. There are four interrelated aspects that concern the environment.[3]

First, to produce sufficient to satisfy domestic and export markets, we must simultaneously remove impediments from past practices, introduce new management procedures that efficiently utilize new technology and genetic resources, and meet the current market demands for quantity, timeliness in delivery, and reliability of quality. To achieve this, the contributions of diverse technologies must be moulded into new management strategies that can respond rapidly to changing circumstance imposed by resource availability, unforeseen constraints and new production objectives.

Second, for the various agricultural products, quality is related partly to nutritional characteristics and partly to prevailing fashion. Some quality characteristics are determined by the production system; others may be affected by storage; and still others may be established, enhanced or degraded during processing. A quality trait has to be redefined not only as something assessed in the product objectively by measurement or subjectively by the senses, but also as something arising out of community attitude to the source or the means of production. Products are of unacceptable quality if they fail on the balance of current prejudice. Sound factual information and better community education might solve the attitudinal problems to quality issues, but there are many examples where misinformation fed by competitors or adversaries has distorted the market.

Third, environmental stewardship, husbandry of resources and welfare of farm animals are matters high in the public concern. There is increasing evidence of public willingness to apply pressure on production and in the market against undesirable practices.

Fourth, there are the continuing issues of global population growth and environmental change. The United Nations *Report on the State of the Environment* expects population to increase by 1.7 billion in the next two decades. Of this increase, 90 per cent will be in countries with little capacity to import food commercially. At the same time global warming will alter world food production capabilities and the distribution of industrial development. The extent of change is unknown and it is unclear in which industries and market sectors Australia will be best able to compete.

However we attempt to look at it, the future is uncertain in terms of population, energy supply, weather, and the potential of science to solve problems. Important issues now are whether agricultural production can be expanded, not just sustained, for the needs of an expanding population, and whether that can be done safely. How well we succeed in achieving a sufficient agriculture will depend heavily on the magnitude of population growth and on decisions made about acceptable levels of use of natural resources and energy in food production.

We could assume that the next doubling of population would require roughly a doubling of food production (it may be somewhat more or less, because many people in the world are at present undernourished while others are oversupplied with food). Where and how agriculture could meet such a challenge are not clear. The principal options are simple—increase yields and/or areas in production. The contribution from more vegetarian diets would be small, because animal production is largely non-competitive with food crops and is offset by the high digestibility of animal products relative to plant products.

Buringh and van Heemst have assessed the production potential of farming without external inputs of fertilizer or fuel.[4] Without energy for irrigation and drainage, the maximum area of land that might be farmed was estimated at 2460 million hectares or 1.7 times the present amount. Grain production proved difficult to estimate, because yields with limiting nutrients vary with the lengths of fallow periods and with amounts of land assigned in rotation to forage legumes. Their conclusion was that even good methods of organic farming (i.e. all wastes and manures recycled plus legume rotations), applied to all possible land, could supply only 3.2 billion tonnes of grain, sufficient for 6.4 billion people. Applied to present farmlands, organic farming would not support the present population of 5 billion. However, an FAO study estimated that 6.5 billion people could be supported on present lands without fertilizer or machines, and that the maximum carrying capacity of the globe when these are used is 33 billion.[5] It seems, then, that a practical upper limit to carrying capacity, with present technology, is something like two doublings of our present population to 20 billion people.[6] Applied to land now being farmed, modern agriculture can probably support a single doubling.

The important conclusion from this study, the basis of which is developed in far more detail by Loomis and Connor, is that fertilizer and other external inputs are essential now and in the future.[7] Therefore we need to identify areas in which intensive farming can be undertaken, and to establish robust systems which use fewer inputs and are more sustainable per unit food produced. At the same time, we must move away from low-input, low-productivity systems in fragile environments where degradation is less readily prevented. Of the world's best farmland in terms of soils and water supply, 60 per cent is found in Asia and North America. Australia's share, 3.4 per cent, is small in absolute terms but is quite large in relation to its population.

The question now is how to address these environmental concerns while proceeding towards an uncertain future. Regardless of directions taken, the basic strategy must be to supply sufficient food for humans through manipulation of environments and plant communities in ways that provide for efficient use of scarce resources. That requires protection from losses during production and afterwards during storage, processing, and distribution. Our view is that current problems can be reduced or eliminated through continued evolution of the existing systems supported by research and education. The view of others, however, is that modern agriculture is bad and must be redirected, perhaps radically, towards 'alternative' methods.

The range of technical, ecological, and social concerns over modern agriculture can be seen in the many 'alternative' groups that espouse changes. Membership of such groups is wide, including farmers and consumers, scientists and non-scientists, conservationists and industrialists. Their motives are equally diverse, ranging from reverence for nature, nostalgia for old ways, dislike of science and technology (particularly of chemistry) and survivalist goals, to political change. Agriculture has always been blessed with theorists of new ways. Cato and Jethro Tull were each important in their time. What seems different now is that more of the criticism and advice directed at agriculture in developed countries is based in political movements that also lack practical experience in farming. A plethora of concepts for farming has emerged as a result, under labels such as 'organic', 'biodynamic', 'alternative', 'appropriate' and 'sustainable'. As is the case with 'organic farming', these concepts are vaguely defined, but usually involve less use of agrichemicals.

Conclusion

Agricultural systems must be sustainable, but they must also be adequately productive to meet society's needs. Modern agriculture has responded to the needs of production and must now respond to the requirement of sustainability. The world population is too large for us to go back to low-input systems. External inputs and sophisticated technology are essential in agriculture. Without them, productivity will spiral downward, leading to poverty and malnutrition for an ever-increasing proportion of the world's population. The real issue for the future is the provision and management of energy, nutrients, and agrichemicals in farming so that needs of society can be met with an acceptable balance between use of land in agriculture and conservation of natural resources. Improvement of agricultural practice is a central challenge. A critical part of this is work towards the reduction of inputs, the replacement of inputs by others with fewer adverse consequences in the system, and particularly the improvement of mixed biological and agrichemical strategies for integration into total farm management. And, needless to say, any such strategies will fail if they do not take into account the social, cultural and political realities of the countries in which they are to be introduced.

Turning Research into Action on the Farm

LEIGH WALTERS AND ALBERT ROVIRA

In recent years there has been a groundswell of concern regarding the sustainability of our agricultural systems. Land degradation, for many people, has become synonymous with poor management of the land, and there is a perception that farmers have not been adopting technology that could reduce the problem. Farmers are faced with the constant dilemma of maintaining profitable enterprises, implementing land management changes, interpreting the plethora of information, and adjusting to market pressures. They require many skills.

Land degradation is an insidious problem that farmers do not want because it threatens their livelihood and the productivity of their land for future generations. Nevertheless, the question remains: why is there a reluctance by many farmers to adopt the research findings which will make the farming systems more sustainable? The answer lies in understanding the technical, financial, social and personal barriers to developing more sustainable systems. This chapter focuses on the challenges for farmers in achieving sustainability, with particular reference to cereal-livestock farming systems. The discussion is relevant to most other agricultural industries, although the emphasis may be different for other systems.

Challenges for Farmers in Applying Sustainable Agricultural Technology

Land degradation, changes in marketing, low profitability due to poor prices, and a growing demand on the farmers' ability to interpret and use information, are all challenges facing our farming communities. The first challenge to farmers is to deal with the nature of Australian soils. Most Australian soils are extremely old because, unlike European, Asian and North American soils, they were not subjected to glaciation and soil renewal. Our soils are low in major nutrients (phosphorus, nitrogen and sulphur), and many are deficient in minor nutrients (molybdenum, manganese, zinc, copper, cobalt, etc.). Superimposed upon these nutrient-poor soils, in many of our cropping areas, are climate and rainfall which are erratic and unreliable, making it difficult for farmers to develop the stable farming systems required to prevent land degradation.

Large advances in agricultural development have occurred through providing solutions to major problems, such as phosphorus deficiency which was remedied by the application of superphosphate in the early 1900s. Trace element research in Australia began in the mid to late 1920s with the identification of manganese deficiency, and the work that followed identified the lack of trace elements in many soils.[1] The correction of these deficiencies culminated in dramatic increases in crop and pasture yields.[2] Nitrogen deficiency was corrected by the introduction of legume pastures into the rotations in the 1950s and 1960s. New cereal varieties have been produced with resistance to a range of diseases. Other problems have also been dealt with as they have been identified. Much of the research dealt with single-factor problems that were relatively easy to identify and correct, and the solutions produced clearly visible results. The production gains and the visual improvements enabled researchers, and later extension officers, to easily demonstrate the benefits of corrective action to farmers.

Australian farmers now compete on the world market against heavily subsidized commodities and yet receive no subsidies themselves. The 1980s and 1990s have been a turbulent period for farmers. The profitability of their industries has eroded to the point that their survival is threatened by even relatively small management errors. In the same period the demands on the farmer have increased dramatically. Farming is no longer a business where a reasonable living can be made with hard work and few business skills. To survive, farmers need to be business-like in their approach and to develop new skills. The changes have not been easy.

Farmers have to take risks to survive. Their ability to compete in highly subsidized world markets has been due to their willingness to adopt new ideas and strive for the most efficient management systems. They have maintained a competitive edge by adopting new technology quickly. Farmers will adopt new technology if they understand the need and consider the change an advantage to their business, and if they feel they can implement the technology effectively at a reasonable cost. If technology is not adopted there is usually a good reason. Understanding the reasons why farmers do not adopt technology is an important function of government advisory services.

The adoption of more sustainable farming systems is a major challenge for rural communities. Often the changes that need to be made are complex and sometimes costly to implement. Understandably, in periods when profits are poor farmers tend to be conservative in their decision-making. Changes usually focus on income-generating activities with a low risk of failure. Even when activities can have a significant commercial benefit in five to ten years time, the immediate cash needs of the business can prevent action being taken. The financial barrier to change should therefore not be confused with a lack of understanding or an unwillingness to alter management. The targeting of research and extension activities has to be based on an understanding of what the barriers are to adopting new technology.

Farmers now have a better understanding of land degradation problems and are recognizing the need to change to improve the sustainability of their farming systems. Farmers have no wish to degrade their land, especially if they understand the consequences. This has created an enormous demand for information and guidance, as indicated by the rapid growth of the Landcare movement.

Responding to the Challenges

Conservation farming

'Conservation farming' is an important aspect of sustainable agriculture. It can be defined as the management of the land in accordance with its capability, the emphasis being on maintaining or enhancing the economic viability of agricultural production, the natural resource base, and other ecosystems which are influenced by agricultural activity.[3] One of the more difficult challenges confronting farmers has been the move to conservation farming. If farms are to remain financially viable, any changes need at least to maintain, and preferably to increase, crop yields and product quality. Otherwise, alternative income has to be generated from other business enterprises.

Conservation farming needs to be integrated into whole farm planning, and to include strategies such as integrated pest and weed control as part of the overall farming system. Conservation farming can take different forms in different districts, but the two essential components are the retention of all stubble and pasture residues and the adoption of minimum or zero tillage (direct drilling). No single recipe for conservation farming can be universally applied. Each paddock on a farm is different and requires its own management approach. If we consider the implications of this statement for the dryland zone of Australia, then the variability of approaches that need to be adopted is staggering. The fact that there is no one exact procedure to suit all situations adds to the complexity of promoting the technology involved in conservation farming.

Although each paddock is unique, the following general principles apply for conservation farming:
1. reduce cultivation and conserve stubble to improve soil quality, reduce erosion and minimize nutrient loss;
2. maintain soil fertility through balanced rotations and the conservative use of fertilizers;
3. where required, use soil ameliorants such as gypsum and lime to improve soil quality;
4. reduce weeds, diseases and pests through crop selection, rotations, resistant varieties, grazing and the strategic use of chemicals.

Adopting a conservation farming approach is not easy, and
has many pitfalls for the unwary. Currently, research provides
solutions for overcoming only some of the barriers to adoption.
There are still many problems that need to be addressed. Table
13.1, which summarizes some of the knowledge required in the
traditional (cultivation) and conservation farming methods,
shows that conservation farming is more complicated and requires a much greater knowledge and understanding.

Table 13.1   Knowledge required in two farming systems

Traditional farming with cultivation	Conservation farming
Machinery maintenance	Machinery maintenance
Tractor driving	Tractor driving
Herbicides—limited	Weeds and herbicides
Superphosphate fertilizer	Diseases, pests and chemicals
Varieties	Varieties
Sowing and harvesting	Sowing and harvesting
Simple rotations	Knowledge of benefits of different rotations—pastures, break crops etc.
	Multiple fertilizers
	Knowledge of soil properties
	Water use efficiency
	Stubble handling

The change to conservation farming is considered highly
desirable because of the benefits from reduced soil and land
degradation. Yet many farmers have not adopted or have only
partially adopted the technology. The reasons for the slow adoption are many and often quite complex. Farmers failing to adopt
new technology are not necessarily rejecting it but are displaying
a caution for which there may be many reasons. The results of
a farmer survey in Victoria shown in Table 13.2 clearly demonstrate that the farmers believe that conservation farming requires expensive equipment they cannot afford.[4] This is linked
to two of the other problems seen as important: 'suitable equipment not available' and 'makes existing equipment redundant'.
However, there are many farmers who have adapted their existing equipment at relatively low cost to handle stubble and to
direct drill.

Table 13.2   Problems with conservation tillage techniques indicated by adopters and non-adopters

Problem	Adopters		Non-adopters	
	Large extent %	Moderate extent %	Small extent %	Not at all %
Suitable equipment not available	29.5	22.2	29.8	30.7
Necessary equipment too costly	76.9	75.3	71.0	66.7
Decreased yield	16.7	32.7	52.4	53.5
Higher costs	20.5	24.1	21.0	28.9
Poor weed control	24.4	36.4	43.5	43.0
Poor control of root diseases	29.5	28.4	37.1	30.7
Unsuitable for soil type	30.8	25.3	37.1	36.8
Stubble management	41.0	38.3	34.7	21.9
Need for nitrogen fertilizer	35.9	38.3	35.5	28.1
Lack of adequate technical advice	26.9	33.3	21.0	28.9
Lack of information	21.8	21.0	21.0	30.7
Makes existing equipment redundant	42.3	44.4	37.9	40.4
Other	10.3	9.3	8.1	7.0
Number of respondents	78	162	124	114

Source: F. T. Hurley, B. C. Fitzgerald, J. T. Harvey and P. P. Oppenheim, *Cropping and Conservation: A Survey of Cultivation Practices in Victorian Grain Growing Areas*, Ballarat College of Advanced Education, Ballarat, Victoria, 1985.

Adopting new technology can sometimes be a compromise. For example, conservation farming may be desirable to reduce erosion, but is it acceptable to increase the use of chemicals to achieve this goal? For farmers in the wheat–sheep belt the increased use of chemicals has been an unavoidable consequence of conservation farming. As benefits accrue from breeding programmes and integrated pest management, cultural and rotation programmes, we would expect the farmer's reliance on chemicals to decline.

The use of herbicides to control weeds requires greater skills than the use of cultivation. Whereas weeds can be controlled by cultivation soon after they emerge, the decisions as to when to spray and with which chemical are not so easy. Weather conditions, whether plants are stressed, the weed species and age are just some of the factors that have to be considered before spraying can commence.

Managing stubble (i.e. remains of crops) as part of conservation farming is more difficult than the use of cultivation and burning to reduce the quantity of stubble to enable the next crop to be sown. Stubble needs to be broken up to encourage adequate decomposition before sowing; otherwise there is a need for specialized stubble-handling sowing equipment. To break up the stubble requires straw choppers, straw and chaff spreaders on harvesters, slashing or mulching of straw, shattering with implements such as prickle chains during the summer, and the use of stock to break up the straw by trampling. Even with these systems, the amount of stubble still remaining at sowing in districts with high rainfall often exceeds the handling capacity of conventional machinery.

Machinery changes involve significant financial commitments so that, even when the farmer is fully committed to the change, machinery has to be upgraded over a number of years. This can be viewed by outsiders as a failure to adopt the technology.

Financing difficulties can affect other work, such as land reclamation and tree planting. Even a modest tree-planting proramme can cost $10 000 a year, yet little return in increased productivity can be expected over a ten-year period. This can be a very significant strain on an average farmer's financial resources, especially when commodity prices are low, although recent changes in the tax law reduce the real cost of conservation practices.

The farmer's first priority is to provide adequate income for the family, and over the past twelve years this has become increasingly difficult. The large debt many farmers are trying to service is evidence of their problem.

Adaptation of technology to the local environment

New technology needs to be adapted to the local environment. For example, in the low-rainfall Mallee areas of Victoria and South Australia the introduction of direct drilling (i.e. sowing without any prior cultivation, a form of conservation farming) resulted in a marked decrease in cereal yields due to a root disease caused by the fungus *Rhizoctonia*. This was not a problem in the high-rainfall areas where direct drilling was largely developed, even though *Rhizoctonia* could be detected there. At present, cultivation and grass control prior to sowing are the two main ways of reducing the severity of the disease. Until alternative strategies can be developed to control the disease in the low-rainfall areas, it will remain a major barrier to the full adoption of direct drilling.

The task of adopting new technology is not always straight-forward. Because of the variation between seasons and soils, farmers need to take considerable care to avoid costly mistakes. The farmers and local advisers therefore interpret, test and modify technology without necessarily relying on an exact 'recipe'. There is an element of gambling when farmers introduce modern technology to their farming systems. To provide additional support to eliminate the need for the gambles would require a major expansion of validation and experimental sites throughout the cropping areas, but this is not possible in the current economic climate where the Departments of Agriculture are being reduced in size.

Research plays a vital role in improving understanding and providing options to overcome barriers to adoption. Previously, a large percentage of the gain in yield from fallowing was attributed to stored moisture; with better understanding we learned that root diseases are reduced by fallowing, due to early grass control in the year prior to the cereal crop. The grasses build up root diseases such as cereal cyst nematode and take-all, and these reduce cereal yields in the following year. The gain from fallowing in many areas with light soil types (e.g. the Mallee) is now considered to be largely due to control of root disease rather than moisture retention. This knowledge has allowed farmers to control grasses by chemical means, thus reducing the risk of erosion and still achieving similar yields. The process of convincing farmers that long-held beliefs were not totally correct was difficult, yet, through an understanding of the reasons behind the increases in yield, benefits of change were realised.

Need for farmers to monitor action and progress

As for any business, farmers need to have achievable and measurable goals. When trying to develop more sustainable farming systems, the goals are not entirely clear or measurable. Little research has been carried out into the ecology of our farming soils, and even less has been translated into information that could be used by farmers. Given limited resources and the plethora of problems to overcome, it is little wonder that our research institutions, advisory services and farmers did not place a greater emphasis on understanding the ecological processes.

There is an urgent need for better measurements of sustainability if farmers are to be able monitor their progress towards the development of sustainable systems. Soil tests answer some of the questions, but they do not at present provide a good guide for the wide range of soil types that exist in our farming areas. Simple field-based measurements are needed to enable farmers to easily measure differences between paddocks and soil types. This would provide farmers with an understanding of what is happening in their soils, and how different farming operations impact on the sustainability of their systems. There is nothing better than direct feedback to motivate a person towards achieving a better result.

At present some of our best guides are yield, grain protein (in wheat) and production potential (yield per millimetre of rainfall in the growing season).[5] These are production, quality and efficiency measures that give an indication of the sustainability of a system if monitored over a number of years. The assumption is that if these measures are increasing, or are at least maintained over time, then the system is moving towards or has achieved sustainability.[6] This conclusion would be reinforced if other soil attributes such as organic matter and nutrient levels were indicating a similar trend. The current indicators are our best guides, given our limited knowledge of the ecosystems.

Non-technical barriers

Not all barriers to adopting new technology are technical or financial. The 'tyranny of distance' can be a barrier to maintaining personal contact with people. In the low-rainfall areas of the Mallee, the large distances between farms and towns make it difficult for farmers to maintain regular contact with their peers and advisers. Telephones and two-way radios have greatly improved communication, but have not eliminated the need for personal contact.

Social or peer-group pressure can be an important factor in influencing a farmer's decisions. Peer pressure can slow down the adoption of new technology if it is viewed as radical by the general community. It can also accelerate the elimination of practices that the community regards as unacceptable. Influencing community opinion on issues such as land degradation can be an important way of facilitating changes on farms. Landcare is a good example of a programme that places peer pressure on farmers in a given catchment or area to 'do the right thing'.

Decisions on farms can be influenced by the family. Unlike many other work situations, especially in large cities, a farmer works and lives in the same environment. The home is the work location; the spouse is usually involved in the business; and socializing usually includes other farmers. The business is a lifestyle. Business decisions, especially major changes, are therefore made with the knowledge, consultation and support of the family. The adoption of new technology can therefore be supported or rejected by the family. The family is usually intimately involved in the business. When new technology is being promoted, it is important to target all the family members that have a role in the operation of the farm. For example, the wife may have a greater affinity for bookkeeping and computers than the husband. To focus only on the husband when trying to encourage better bookkeeping practices or the use of computers may miss an important and influential audience.

Farmers need to be multi-skilled

Many skills are required by farmers. Few jobs require a person to be not only an agronomist, but also a stock specialist, manager, bookkeeper, mechanic, marketer, handyman and soil conservationist. Add to this list the additional planning skills required for managing more complex systems and introducing major changes, the growing pressure to become computer-literate, and the necessity to develop trading skills as more goods are traded on a deregulated market. Trading alone requires new information sources, an understanding of the markets and an ability to obtain timely information. The need for timely information has resulted in the use of facsimile machines as a way of keeping up with market prices. Changes such as these may seem acceptable to many people, but for farmers who are trying to manage a farm they can become just another task on a long list of needs.

Many of the skills a farmer develops are learned from parents, interactions with other farmers, field days, local training workshops or courses, reading and the experience of managing a farm business. Most farmers do not have post-secondary education such as an agricultural degree or diploma. Formal schooling therefore provides farmers with only basic skills. Any other skills that are needed have to be learnt informally (e.g. field days) or through short courses provided by TAFE and other educational programmes. Conservation farming requires good technical, planning and financial management skills; since many farmers lack some of these skills, it is not surprising that they are having difficulty in adopting the technology.

Considering what is being asked of them, farmers have introduced changes relatively quickly. As with any group of people, though, change comes more easily for some than others.

Helping Farmers Respond to the Challenges

Farmers are being saturated with information from a wide range of sources. A common complaint farmers make is that they do not have the time to read all the publications. This raises two issues of concern for government agencies introducing new technology.

First, new ideas can be considered only if people know about them. If farmers are having difficulty identifying good information because of the volume of publications they have to peruse, then the approach to providing information may need to change. New technology may need to be marketed like other commercial products in order to attract the attention of the farmer. The risk in this approach is that it is seen as another commercial product competing for the farmer's attention.

Government departments also promote a wide range of technologies. If widely used, intensive marketing of technology would also be very expensive.

Second, if farmers have so little time to read, then how are they tackling the planning aspects of management? This is an extremely important point, as many of the changes towards more sustainable systems require considerable planning if they are to be adopted successfully. Unfortunately, current indications are that farmers are not coping well with planning. Poor office and financial management skills are hampering the planning process. This means that what can be expected from farmers may need to change. If decisions are based on largely short-term imperatives, then the focus on long-term goals will be an extremely difficult concept to sell except in a very general sense.

Information overload is a problem not only for farmers. The quantity of information is also causing problems for advisory services. With increased workloads it is becoming more difficult for advisers to read all the appropriate publications. As the problem worsens, there is the chance that new technology will not be utilized because the adviser is not aware of its existence. Current cuts in state government funds for agricultural extension staff further exacerbate the problem of technology transfer. This may be partly resolved through the employment of paid consultants by the farmers, and the greater involvement of the federal government through industry research and development corporations.

Developing sustainable farming systems is not easy. However, many rural communities are responding positively to the challenge. The enormous support for Landcare activities has resulted in a rapid expansion of groups throughout Australia. Other group-based programmes such as SoilCare (northern Victoria), Farm Management 500 (central Victoria and southern New South Wales), Farm Advance (central Victoria), Wimmera Conservation Farming Association (Victorian Wimmera), Farm Trees and Right Rotations (South Australia), to name a few, are all targeted at developing more sustainable farming systems. They represent a new wave of action driven from the grassroots upwards rather than from the researcher downwards.

In Figure 13.1 two models for information transfer are illustrated. In the linear model (A) the client (farmer) is simply the recipient of the information and has no involvement in generating it. Current thinking favours model B in which the farmer, researcher and extension officer (or consultant) work through the problem together so that the farmer has a sense of ownership in the solution and is more likely to adopt the technology. With the present reduction in government-provided extension services, the employment of private consultants becomes necessary; this places a financial load (and possibly a barrier to adoption) on the farmer.

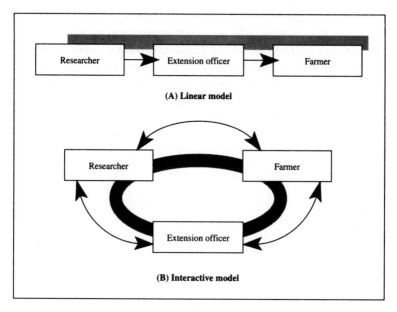

Figure 13.1   Two models of transfer of information to farmers

Conclusion

When considering conservation farming strategies, we need to remember the complexity of the issues. We are seeking to implement major changes. This requires awareness, training and supporting information. This process needs to occur with researchers, advisers, farmers and the general community. The support and understanding by the general community of the challenges being faced by the farmers is an important part of the process. It is counterproductive for rural communities to keep on hearing of the damage to our land through agriculture without acknowledgment of the enormous effort going towards developing more sustainable systems. People respond to positive encouragement. There is a lot happening; farmers are moving in the right direction, but change will not happen overnight. What is needed by farmers most of all is encouragement for their efforts, achievable goals and support to overcome the barriers to change.

The National Farmers Federation and the Australian Conservation Foundation worked together with the federal government to set up the Decade of Landcare. This co-operation will greatly encourage the farmers in their pursuit of sustainable farming systems. At last farmers will feel that they have the support of the urban population in this difficult task of improving productivity and sustainability.

14

The View from the Farm

PETER SMALL

The economic well-being of Australian society has been, and will remain, closely linked to the fortunes of the agricultural sector. Forty or fifty years ago the interdependence of the city and country was well understood, particularly by city people, many of whom had strong links to the country through family and friends. However, although we live in an age of rapidly improving communications, the gulf between those who live in our large cities and those who live and work in our rural communities is ever widening.

Although many city people in this country have a picture of inefficient and outdated farming systems and declining soil fertility, the true situation is definitely to the contrary. The improvement in the quality of our soils, the control of soil erosion, and the associated increased productivity of our crops and pastures would not have been believed possible at the beginning of this century. That is not to suggest we cannot do a lot better; we can. But what is needed to achieve an even higher standard, and ensure the long-term sustainability of our nation's soil, water and air?

Sustainability

Sustainability is a buzz-word that emerged in the 1980s. Environmental sustainability was the concept, recognized and supported by both farmers and greens. As we settle into the economic realities of the 1990s, though, farmers are becoming increasingly preoccupied with financial sustainability; in short, survival. To the farmer, sustainability means conserving and improving resources: the land, the water, and the means of production. But above all the farmer must maintain financial viability, through droughts, depressions and like calamities.

Farmers who are unable to generate profits cannot, with the best will in the world, do much about land degradation. Without financial viability there is no future; in time the farming family becomes just another statistic, like thousands that have left before them, a statistic registered with the Rural Adjustment Scheme. Generations of toil, and in many cases deprivation, are lost forever.

Most farmers are conservationists. They know that their future depends on being so. But despite having this as the long-term goal, there are often short-term pressures that lead to deferral of action. How important is that deferral? How extreme will be the cumulative effects of a period of continuous imperceptible change? How can they be measured by the landowner or the land manager?

Farmers are subject to the same frailties as their fellow beings. All farmers find it hard to maintain good farming practices in times of emotional stress. Emotional stress manifests itself in lack of motivation. Emotional stress can be precipitated by the traumas of drought, bushfires, and financial depression, or by personal or family crisis of one sort or another. Studies in New Zealand show that lack of motivation, and of financial and business skills, are far more important causes of farm failure than actual levels of debt.[1]

While environmentalists who survey agriculture from the security of a salaried position can often produce evidence to denigrate farmers' attempts at conservation, they rarely seek to understand the situation from the farmer's point of view. It is important to understand that farmers have very good reasons for their actions. Their decisions are based on their economic predicament, their knowledge, and their understanding of the issues unique to their situation. Knowledge is acquired by careful observation of their environment, and by learning through a range of formal and informal media. However, there are many gaps in our knowledge of these complex issues, and farmers naturally reflect in a practical way the implications of this lack of knowledge. Many of our leading farmers have all the intellectual attributes of leading scientists, but with one serious deficiency: they rarely record their observations. And their failure to measure and record is one of our fundamental problems! (The development of the Australian merino is probably the supreme example of a largely unrecorded success story of animal science. The history of this process has been recorded recently by Massy.)[2]

Key Environmental Problems Facing Farmers

Although I spent some time in my earlier years gaining farming experience in Queensland, my own forebears initially selected land in the Victorian Mallee in the 1870s. My father sold this land when I was twelve years old. We then farmed in the southern Wimmera on the northern slopes of the Great Divide. In 1976 my wife and I purchased land on the edge of the Dundas Tableland in the Western District of Victoria. One way or another I have kept contact with a cross-section of Australian agriculture.

With my background, what concerns me most are the perceptions and gaps in understanding of those who, unlike farmers, are not responsible for producing clean, safe food at reasonable cost on a sustainable basis. Recently I spent some time in southeastern Queensland. Kangaroos are in plague proportions. Farmers' killing quotas have been exhausted. An animal that produces high-quality protein and some of the softest and most supple leather in the world is being left to rot as farmers battle to save their crops. The dingo is encroaching further and further into sheep areas with devastating results. Noxious weeds are becoming more and more difficult to control because of diminishing labour resources. Wild pigs are breeding up into unbelievable numbers. (What if there is an outbreak of an exotic disease?) Across Australia, the rabbit is breeding up again to pre-myxomatosis numbers. The feral cat is on the increase, both in size and numbers. There are wild goats, flocks of corellas and cockatoos in plague proportions, and so the list goes on. It sounds familiar: 'We'll all be rooned, said Hanrahan'.

For most farmers, the control of vermin and noxious weeds is a continual battle. This problem is often exacerbated where national parks are created or grazing rights are removed from Crown lands. My observation is that when pastoral country is destocked and converted into national parks there is, after a few years, less natural herbage and increased noxious weeds. Governments are not very effective at controlling introduced and native species. As European governments have found, it is more effective to pay the farmer to tend the countryside than to displace farming families and pay public servants to do the task.

Environmental Problems Facing Southwestern Victorian Farmers

The most important environmental problems facing agriculture in southwestern Victoria are rising water tables and the associated problem of salinity, together with increasing soil acidity. The salinity in the Murray–Darling system is accentuated by the extensive inland irrigation system. In southwestern Victoria removal of trees is blamed for the rising water tables and associated salinity problems. Indeed, lack of trees has also been blamed for the decline in soil fertility and increasing soil acidification.[3] Across Australia farmers, observing tree decline and salinity, have embraced the Landcare movement. This strongly promoted movement is probably as active in my own district as anywhere in Australia. The almost frantic planting of trees in the late 1980s has dramatically surpassed the massive planting of sugar gums for stock shelter across the Western District plains during the 1920s.

No-one will disagree about the value of trees for stock shelter, for maintaining the aesthetics of the landscape, or for hiding areas of serious land degradation. However, their capacity to turn the tide on rising water tables, salinity and increasing soil acidity is another matter. The value of trees in solving these complex problems has been oversold. Many farmers investing scarce resources (albeit in more prosperous times) in trees, as a solution to salting, are now having second thoughts. To a large extent they have been sold a pup. Much of the 'grow more trees' campaign has been based on 'best bet' scenarios, rather than real scientific evidence.[4]

In their book *Greening a Brown Land*, Barr and Cary postulate, 'The loss of the trees has become a symbol of the cause of land degradation ... tree planting is symbolic, rather than a real activity for land holders in combating land degradation'.[5] In the wider sphere of community belief, forests and trees also have this symbolic value. Symbolic beliefs are often necessary for community consensus so that agreed community activity may occur. But it is necessary to see beyond the symbols; we must rationally establish the causes of land degradation problems in order to establish rational and feasible solutions.

As we approach the end of this century and the end of this millennium, planting of trees has become a symbol not only of combating land degradation, but also of the decline of many of our institutions that are failing to withstand the pressures of change. Planting of trees is a positive thing we can all do that is symbolic of our desire for a better, sustainable world. However, we need to be careful in the way we use our scarce resources. Planting of trees should not become a panacea, and distract us from making the difficult decisions to secure the future. The analogy is the battle flag. The flag is the rallying point, a morale booster. But even if we all gather around the flag and defend and support it, this will not win the war, nor even contribute much to the winning.

While there is a certain nostalgic attraction about looking back to the way it all was before the Europeans began farming Australia, really there is no going back. What we can do is to learn from the past and use our skill and ingenuity to build a better future.

In southwestern Victoria we are not even sure that going back to the past would bring us the stability to our land system we so desperately desire. While white settlement has obviously accelerated the process of land degradation, much may well have occurred even if we had stayed away. When Major Mitchell first moved across the Western District there was evidence of high water tables, and early settlers described land slips and similar land degradation problems. In any event we are here, and no-one knows how dense a forest we would need to regain the equilibrium, even if that was possible or desirable. We are farmers, not foresters. While the important reafforestation programmes must continue, for the bulk of our farmlands Australia needs the wealth we can generate now, not in forty or fifty years time. We have no alternative but to move on.

Marshalling the Energy, Knowledge and Resources

The first essential, then, is for a common community attitude to embrace the future. The symbolism of planting trees to generate new growth needs to be translated into community action for growth and wealth creation. It is vital that the city and the country be locked together in this endeavour; the self-interest of both groups must be seen to be entwined; the city needs to appreciate that clean, high-quality food, water and fibre come at a price, and that good land and water protection in a long-term sustainable system also comes at a price. The price is not just in the cost of the product; we must also pay for a sophisticated system of research and development, education and training. The procedures need to be in place so that, having acquired the knowledge and the skill, the farmer has access to low-interest loans, and in particular to taxation incentives to encourage investments in sustainable systems, particularly in times of high commodity prices.

A change in community attitude is needed to nurture and reward keen young scientists so that good relevant science can proceed. (The community attitudes of the last five to ten years have not only aborted much valuable research and development; they have also disenfranchised many active research workers with agile minds, people we desperately need to work on these complex problems.)

Individual landholders need the support of better methods of analysis of the state of their land; of better targeted intervention methods; of ongoing monitoring procedures; of short-term advantage, rather than disadvantage arising from financial arrangements for putting their management practices to work. This, in the final analysis, requires that a dominant urban population understands the need to give support to landholders in best management practice, much of which will be very indirect.

We need to marshal our resources aggressively in order to seek real solutions to some of the serious gaps in our knowledge about land degradation. We need an end to the promotion of the 'warm inside feeling' solutions, based on political expediency. There is a need for decision-makers to leave their comfort zone and develop closer links with industry. A partnership between industry, research workers and educationalists will form a powerful combination to solve these complex problems. As Professor Egan said in his presidential address to the Australian Society of Animal Production in 1992, 'Above all, scientists and agriculturalists ... have expertise in the solution of complex problems and are well placed collectively to apply problems of wider scope. They are, by and large optimistic, but keenly aware of the perils of self-delusion'.[6]

Marshalling of resources, both physical and intellectual, will not be easy, but it can be done. We must face the chilling reality that to fail will have immense implications. Not only does our nation need a land system that is sustainable in the long term, producing food and fibre for our urban population, but we urgently need greatly increased export income to halt our ballooning external debt. For a business burdened with debt there are only three ways out: increased productivity, asset sales, or bankruptcy. For a nation it is no different. Our national survival is at stake.

Conclusion

I have not been able to canvass all the environmental issues arising for farmers in the quest for sustainable agricultural production; there are many more. What I have tried to do is suggest that together we explore a different road. That road could be built on the consensus which is symbolized by the tree. What is needed is action: concerted action by government, industry, scientists, educationalists and environmentalists. We need a new vision for Australia, based on the mutual self-interest of city and country entwined together with a strong emphasis on the generation of new wealth, so that we can afford a sustainable system and escape the net of poverty and despondency.

15

Alternatives for Achieving Sustainable Land Use

JOHN BRADSEN

Previous chapters in this part have examined degradation of farming and pastoral lands from the perspective of the farmer and agriculturalist. Change to a more sustainable agriculture will be achieved only if the farming industry and agricultural researchers associated with it continue their efforts to develop, disseminate and introduce new practices. We cannot, however, assume that these efforts will be successful on their own. The results of industry initiative or self-regulation alone have been unpromising, however much they are to be encouraged.

This chapter explores what other measures might be taken to bring about environmental change. Before discussing some specific examples, it reviews the nature of sustainability. The questions that it poses are: What does sustainability mean and how should it be interpreted when applied to farming and pastoral lands? If sustainability is the goal, what, then, are the techniques of thought and social organization that will help us achieve it? To explore these questions, the chapter reviews the concept of biodiversity and the roles of science, ethics, economics and the law. The final part of the chapter reviews the operation of Landcare in Australia, and identifies its achievements and shortcomings. The conclusion is drawn that the shortcomings of such programmes can be dealt with only by using the law. Some examples of successful and unsuccessful South Australian legislation are used to support this conclusion.

Sustainability

Where land is being blatantly abused it is usually clear that its use is unsustainable. But sustainability is generally a far more insidious question. It must be asked whether the community has really begun to come to terms with its true meaning and effect.

It is, at times, still suggested that sustainable development means sustaining development. It is common to read that there is something called a balance between sustaining natural resources and development. It is often said that we cannot have a sustained environment without development. The truth is that any period of development is a transient cultural phase.[1] The economy is not fundamental. The earth is. The environment sets the limits and we have to endeavour to work out what the sustainable limits are.

One of our greatest problems is to be realistic. This is particularly difficult for modern Western communities caught up in a constant whirl of change, enthralled by technology and constantly assailed by immediate political, economic and environmental issues. It is difficult, too, for communities which are consumed with dominating and manipulating nature, and which have come virtually to worship themselves and their cleverness, to accept that the natural world really does set limits. And it is difficult to know what they are. Realism is readily decried as pessimism. Because change is the norm, we are told to be optimistic. But both pessimism and optimism are merely states of unreality.

An associated problem is an apparent determination not to learn from the past. The increasing information about the archaeology of land degradation would appear to issue a stern warning. The evidence suggests that no civilization, that is a society based upon the intensive use of agricultural land to support large urban populations, has been sustainable. It shows that a number have collapsed, and suggests that land used too intensively may have a life expectancy of only several hundred years.[2]

In assessing our civilization we can see that we are placing more intensive and widespread demands upon natural systems, especially the land, than has ever occurred before, with land degradation reflecting an increasingly intense relationship between land and cities. In short, the present period of civilization, which is the most disruptive of natural systems that the earth has known, may also have to be the first to avoid such a collapse.

Fundamental Natural Resources

The focus in this book is on land. It is necessary, however, to bear in mind all four fundamental natural resources, air, water, land and biodiversity (that is the balance of life in all its diversity), since they are all inextricably interrelated. This is best explained by reference to the most fundamental, the unique feature of earth: its biodiversity.

Biodiversity is crucial, for example, in having created and in maintaining earth's air; it purifies surface water and acts as a pump to hold down damaging saline groundwater; and land, as soil, is the continuing product of life, from deep-rooted perennials through worms to microbes. Biodiversity provides not just particular species, but, more importantly maintains in balance the health of the earth's natural systems. Indeed it is suggested that conservation of biodiversity will increasingly be seen as the litmus test for the sustainable use, or sufficient conservation, of the other resources.

The Scandinavians, for example, have taken care to develop sustainable-use policies for their native forests. For some time it was thought that their timber harvesting was sustainable with long rotations and no loss of soil or nutrients. But with the shift from complex natural forest to simpler, more monocultural structures, species are under threat and ecosystems are showing signs of stress. The whole concept of sustainable use is being reconsidered. The bottom line is no longer the sustainable production of timber but the sustaining of biodiversity. There is a direct parallel, for example, with the use of Australia's arid zone or New Zealand's high country.

Similar comments can be made about the sustainable use of cleared, productive land. To date we have adopted various physical or chemical measures of degradation, including whether soil is being physically lost or its structure damaged, or whether it is becoming acidic, saline, non-wetting, deficient in nutrients, and so on. But these measures of degradation are somewhat crude, and in terms of true sustainability it is suggested that the test of degradation or sustainability will be the life in the soil. Perhaps the most obvious example is the humble worm. Its importance can be seen in research which shows that it can lift productivity. It can be seen too in the damage caused in Europe by the feral New Zealand flat worm. Similarly, algae blooms may be better understood and dealt with in terms of biodiversity than in terms of nutrient overload.

Biodiversity, now being increasingly recognized as an issue involving the health of ecosystems, even the processes of evolution itself is, of course, so all-encompassing that it is difficult to comprehend. In order to reduce it to a more graspable level, the focus tends to be on threatened species and rates of extinction. Although rather simplistic, this does give some sense of the well-being and functioning of ecosystems. Extinction is, of course, part of the evolution of life; thus to speak of stopping extinction is meaningless. In this regard, therefore, conservation of biodiversity is largely concerned with identifying acceptable rates of change.

Although we still do not know how many species inhabit the earth, respected biologists, using the best data available, suggest that up to one-quarter of all species on earth may be lost within several decades.[3] This has been described as a rate of extinction some ten thousand times faster than previous mass extinctions. But this, too, is a very elusive concept. In more immediate terms we can say that it would appear that one in every eight species of birds in Australia is now under threat,[4] as are numerous plant species, while Australia's record in destroying smaller mammal species is the world's worst.

Having said these things, complex questions about the significance of extinction or reduced conservation status of species still remain to be explored. For present purposes, suffice it to say that there are both practical reasons for conserving species, including the utility provided by them and their ecosystems, and moral reasons, including respect for other forms of life.[5]

It also remains to be understood how the conservation status of species indicates whether a sufficient level of biodiversity conservation is being achieved. Presumably the great majority of species should be conserved in a non-threatened, even non-rare, state as an indicator of ecosystem health, if not for other reasons. The point of conserving species in a threatened state, whether in cages or genetically captive in isolated pockets of vegetation, is questionable. In general, the more we learn, the more we see that larger areas and less disturbance are required to sustain complex ecosystems that are rich in species.

Techniques of Thought and Social Organization

It is those working the land in agriculture who are most directly involved in coming to terms with the issues of both land degradation and biodiversity. But it is above all the burgeoning cities which are pushing the land so hard. They must come to understand and be prepared to assist. Both need to grapple with the techniques of thought and social organization which are crucial.

There are various techniques and modes of thought used to understand and organize the world, including, in particular, our use of natural resources. The focus below is on science, attitude and ethics, economics and law.

Science

Science, along with its associated technology, is increasingly replacing myth as central to the Western understanding of the world. It is now regularly called on for help where the conservation of natural resources is concerned. Indeed, if one explores environmental legislation, it is perhaps most particularly its reliance on science which gives it a distinctive flavour. Most law is based upon certain experience and assumptions about human nature. Environmental law adds these perspectives as they apply to the natural world.

Science has, however, generated some myths of its own. There tends to be a belief at all levels of society that science can 'prove' things and tell us when and how to act. This view, and the associated one that technology can solve all our problems, is not science but faith or myth. Science cannot 'prove' things, especially in many areas relating to biology. All it can do is to indicate probabilities or degrees of likelihood. And having done that, science cannot tell the community what level of risk it should take.

Faith in the notion of scientific proof typically leads people to await something definitive, or at least proof 'beyond reasonable doubt', to use the criminal-law concept of proof. Increasingly it is appreciated, however, that science needs to present information in graded terms, including proof in civil-law terms of 'more likely than not'. The precautionary principle suggests that society should consider taking action in response to lower degrees of probability.

In short, the relevant degree of proof and the level of acceptable risk, while guided by science, are matters of judgement for ordinary well-informed people of good sense. This is not to decry science, despite the limits inherent in reductionist thinking, especially in ecological matters. It is crucial both to understand and to deal with environmental issues. But science itself needs to be better presented and understood.[6]

Attitudes and ethics

It is frequently stated that to ensure environmental sustainability we must change people's attitudes.[7] In the context of land it is often said that we must develop a land conservation ethic, a view sought to be hardened by adding the concept of commitment.

There is no doubt that everyone, including so-called purely practical people, have a thick, often unexamined, even unknown, underlayer of attitudes, ethics, theories, myths, morality, commitments and so on, and that these are associated with their behaviour. (These notions all differ but tend to be muddled together. In any event the issue in hand has been discussed largely in terms of attitude.) The question is whether effective (that is, sufficient, comprehensive, sustained) conservation can be achieved through seeking to change attitudes.

The link between attitudes and behaviour is far from clear. We now have an upsurge of environmental concern and we see some impressive responses. The link may be there, particularly where issues are relatively simple, like concern for whales or even planting trees. But an assumption that, by changing people's attitudes, a major and sustained change can be effected in their behaviour in relation to complex environmental issues like preventing land degradation or conserving biodiversity is far from well made out.

There are other significant circumstances which illustrate that it is simply not possible to predict behaviour from attitude. For instance, a Christian attitude embraces people from Mother Teresa to the protagonists in Northern Ireland. In the secular sphere there is probably no stronger attitude or ethic than free enterprise. Yet this has typically been accompanied by anti-competitive behaviour by many of its proponents who have objected to trade-practices legislation which defines key elements of free enterprise and requires adherence to them.

In the context of land management, history reveals that attempts have been made to encourage sustainable use through emphasis on attitudes. Following the creation of the Dust Bowl, the United States embarked on what was perhaps the longest-running, most intensive effort ever undertaken to procure an outcome by this means. The cost was huge, some $30 million.[8] Yet it was assessed in the 1980s as between not very effective and a colossal failure.[9]

Research specifically on the point suggests that there is no clear link between attitudes and ethics and behaviour.[10] The issue can be thought of in terms of the common statement that farmers are natural conservationists. (If this were so why the stated need for education, extension and changed attitudes?) The issue is far more complex. The explanation of the complexity lies in part in the conceptualizations of attitudes and ethics. A land-use attitude or ethic may be a highly complex mixture of hopes, fears, good intentions, altruism and self-interest, with much scope for self-contradiction and self-deception, all permeated with things like Western economic and biblical concepts of resource use, and faith in technology. Thus to expect that a so-called change in attitude or ethic can change behaviour and thereby achieve effective (sufficient, comprehensive, sustained) land conservation is hardly realistic. If one subjects the added concept of voluntary commitment to similar scrutiny, one is left wondering how much more it can be relied upon than a new year's resolution.

Rather than begin with assumptions about the role of attitudes, the discussion should be focused directly on the important issue, which is long-term behaviour. It is not suggested that it does not matter what a person's attitude is, and that all that matters is behaviour. Society needs to be congenial as well as sustainable. But the focus on behaviour places particular emphasis on the two primary techniques of social organization, economics and law.

Economics

Economics and law both embody their own modes of thought, but they are more particularly the two pre-eminent modes of social or community organization.

Considerable benefits flow from organizing society through free-market economics. There is no doubting its capacity to generate economic growth and wealth. Often overlooked, but perhaps even more significant, is its role in a system of political and social freedom. But with regard to natural resources, it has limitations in favouring the individual and the short-term over the community and the long-term. There is no doubt that economics has not taken adequately into account the slow, insidious process of land degradation or the destruction of species and ecosystems.[11]

Inadequate conservation is well illustrated by two tenets of economics. First, it has often assumed that markets are perfect. The more realistic assumption about markets in the land and biodiversity conservation context is that they are imperfect. At best the question should be regarded as open, in which case the balance will swing towards imperfection for a number of reasons, including externalities. Most fundamentally, while markets have value in allocating within an economic system, they are very poor at determining outer limits to the system.

Second, economics relies heavily on cost-benefit analysis. Central to this technique is the discount rate, which discounts future benefits and sets a time-frame for the analysis. The problem is that any rate at which the future is discounted assumes that the resource can be used up or that at some future point benefits stop. This is untenable. Land and ecosystems must go on providing benefits indefinitely.

The economic view of the world not only makes certain strained assumptions about human nature and rests heavily on faith in technology, but it appears also to assume that economics, like the natural world, is fundamental. However, in considerable measure economics is a cultural construct in that many of the economic parameters such as individual properties are artificial constructs, in terms of factors like their size, productive capacity and asset structure. This can be expressed in another way by saying that it is not possible to insist that individual properties are both profitable and sustainable. And this is true at the level of individuals or nations. At a global level, and ultimately that is the key level in terms of sustainability, the proposition is meaningless.[12]

The history of Australian land degradation is very much one of bouts of intensive settlement driven by government and community, followed by property consolidation and asset-base reconstruction through the forces of bankruptcy and degradation. To insist upon the need for both profitability and conservation in this climate is to invite more of the same. To insist that economics and natural resources are equally fundamental is false. Ultimately the fundamental element is the resource. If it is ruined, society loses both the resource and the profit. It is not possible to insist on both.

The techniques for using economic incentives may be as much an exercise in law as in economics. Moreover, as Chisholm illustrates, they can be very complex.[13] They may have a role where resources are physically 'in common', and where targeting is unimportant, but are of limited use where this is not the case. In any event it is questionable whether society should subsidize any group to use society's resources on a sustained basis.

Law

When the idea of using law is invoked in the natural resource context, many people immediately assume that this means 'regulation'; it is further assumed that this interferes with private property as if that were inherently wrong. These two issues, private property and regulation, require comment.[14]

Private property is an institution (like the institution of free markets assisted by the law of private contract) that is important to individual, social and political freedom and to the collection and protection of individual wealth. But it is above all a community institution like democracy, participated in by individuals but adopted and protected by the community ultimately for its benefit as a whole.

It is illogical to argue, therefore, that private property rights can prevail over the interests of the community at large. The history of private property shows that the community's compelling concerns will prevail. The critical question, therefore, is not whether there is an interference with private property, but whether there is a compelling community interest. Private property rights do not include the right to destroy land or threaten the existence of other forms of life.

The concept of 'regulation', which is typically assumed to involve negative, punitive, criminal law, appears to stand in the way of constructive discussion about the use of law more than any other issue. It is true that law may need to have that operation in some instances. But legislation is a very flexible instrument, with its design limited only by the imagination. Far more emphasis should be placed on the unique capacity of law as a community instrument for realistically assessing, facilitating or organizing, particularly where long-term issues are involved.

In the long history of land-conservation legislation, several legal models have been used. One is the enactment of laws in the traditional legal mould which prohibit aspects of land degradation, with an infringement attracting a criminal penalty. This is the sort of law which is typically referred to as 'regulation'. There may be a place for some such laws in the land conservation context, such as protecting a person who has a right of entry and inspection. But as a technique for ensuring the ongoing better management of land such laws are almost entirely inappropriate, although they may be on the increase in the United States.[15] Laws which leave land conservation optional, relying on discretionary action, education, extension and incentives have been tried at great length. They do not have a good record of effectiveness.[16]

Another approach has been to seek to prevent land degradation as part of the system for protecting water quality. It is one of the oddities of history that the philosophical difficulty in accepting restraints on land use to protect the land have tended to vanish when much the same restraints are applied to protect water quality from the effect of land use. Yet damage to water may be more repairable than damage to land.

The most recent United States technique was initiated in the 1985 Farm Bill and extended in subsequent bills. The major features were the Conservation Reserve Program and cost compliance. Under the former, some forty-five million acres of land were taken out of commercial production. Landholders were compensated for not cultivating the land, although the restriction lasted for only ten years and the land in question had an average annual soil loss of about twenty-nine tons per acre. That is, landholders were compensated for not doing what, on any test of sustainability, they should not have been doing in the first place.

Cost compliance is the more potent part of the scheme. It relies on the fact that farmers get such generous agricultural subsidies that many, perhaps most, landholders could not prosper without them. The law provides that the landholders receive such subsidies only if they comply with a property plan which they must prepare and which addresses the question of land conservation. The concept is not unlike a fine; the differences are that the law does not formally prohibit conduct (though this is morally implicit), and that the economic impact is probably greater than could be expected of a fine for similar conduct.

The same requirement applies to the use of land where the Conservation Reserve Program has expired, and to certain forms of development of land which may have environmental consequences. Indeed, the latest farm bill goes well beyond traditional soil conservation and includes a range of environmental consequences of farming, including impacts on wildlife. This will undoubtedly be extended as biodiversity becomes an increasingly significant issue. These property plans had to be prepared by 1992, and there was a flurry of activity in order to comply.[17]

In Australia the most interesting and perhaps effective use of law is found in South Australia. This will be examined after looking at the most dominant Australian land conservation movement, known as Landcare.

Landcare

Landcare is typically described as a community-driven movement supported with government assistance. It is a remarkable and apparently uniquely Australian institution, one not easy to define.[18] Its present form can be traced back to an agreement between the National Farmers Federation and the Australian Conservation Foundation to present a joint package to the Commonwealth. The result was the 1990s Decade of Landcare during which the Commonwealth promised to make available more than $300 million for Landcare. The Commonwealth and each state prepared a Landcare plan.[19] Groups of all shapes and sizes have formulated Landcare projects and applied for assistance. These applications are assessed and may be funded. Numerous groups do not incorporate in order to seek funds, but operate on a purely voluntary basis.

There is no doubt that Landcare is a remarkable phenomenon, with the development of some thousand Landcare groups. Although much good work is being done, problems remain. These are discussed below.

Landholder care

Landcare is not associated with sufficient concern for landholder care. The long history of government-driven closer settlement followed by property consolidation through the combined forces of degradation and bankruptcy is not yet worked through. It is taking on a new significance as the full implications of sustainability sink in. It is in the interests of both landholder care and land care that a fresh emphasis should be given to viability and reconstruction.

Viability is, however, not a simple question. Landholders may be non-viable for various reasons, including international forces, government policies or unwise business decisions. Concerning the latter it must be recognized that the history of land use in Australia has been concerned not just with production but also with speculation. For some the land is a way of life, but for many it is a business, including an element of speculation. Low returns on capital may reflect either.

Where landholders have made unwise business decisions, it is difficult to justify support. Nevertheless many genuine landholders for legitimate reasons find themselves with properties which in the long term are non-viable. They cannot be expected to manage on a sustainable basis. A comprehensive land care package should, therefore, include consideration of the structure of rural Australia. This must include landholder care which cannot be provided by Landcare.

Multiplicity of programmes

Landcare is criticized as being too confusing because of its overlap with a range of other programmes, including the National Soil Conservation Programme,[20] One Billion Trees, Greening Australia, Save the Bush, the Murray–Darling Basin Commission activities[21] and now the biodiversity strategy.[22] There is substance in this concern. The difficulty lies not so much in the multiplicity of programmes, but rather in the lack of rigour in their development. Simply cobbling the programmes together is not the answer.

These programmes appear to have a number of aspects. One is a concern with fundamental questions about the use and management of land and the maintenance of biodiversity, though it must be said that they appear to receive too little attention. Another is an enthusiasm for tree planting. Landcare embraces many activities, but in good measure is a tree planting programme. The rash of trees may not play a major role in achieving biodiversity. The point is that greater consideration needs to be given to the goals sought to be achieved and that programmes need to be structured accordingly. In short, confusion seems to lie more in uncertain goals than in too many programmes.

This point may be illustrated by the example of the distinction between tree planting and retention of vegetation for biodiversity conservation. Simply in terms of numbers of trees in the ground, saving existing vegetation provides far greater value. In South Australia alone, for example, the vegetation clearance legislation has seen over half a billion trees and probably over twenty-five billion understorey plants of significant size saved from clearance in about six years. The greatest biodiversity need in Australia is to preclude further clearance until its biological significance has been properly assessed. The ecosystems associated with original vegetation are virtually impossible to replicate, and a number are rare or threatened. Moreover, many of Australia's threatened plants are not trees but understorey plants.

The planting or natural regeneration of vegetation for species conservation requires a set of criteria not necessarily included in general tree planting, including using appropriate genetic material and looking at appropriate vegetation linkages.

Accountability

A further difficulty is that Landcare money may be distributed with insufficient accountability. This does not refer to the detailed recording of expenditure but rather to the broader thrust of Landcare. When money is sought, programmes are assessed and placed in order of priority against criteria. But there is insufficient assurance that these funds will be directed to activities which are truly sustainable. As a leading Farmers' Federation figure once said, Landcare is in danger of becoming a honeypot. In the United States, reviews of assistance provided for land conservation suggest that funds have tended increasingly to go to enhancing shorter-term production and other goals rather than sustaining long-term productivity. The assistance became a pork barrel.

Accountability includes ensuring that Landcare gains are not unwisely undone. In the US many millions of trees, planted at public expense in the huge tree planting effort which followed the Dust Bowl, were later dealt with according to short-term private interests. In Australia land conservation works undertaken with public assistance have been undone to improve short-term profitability. The question is whether there is sufficient assurance that Landcare work undertaken with community assistance will be effectively secured in accordance with the needs of sustainable use.

Effectiveness

Those who extol Landcare's virtues point out that Landcare groups embrace up to 30 per cent of landholders.[23] This is undoubtedly an achievement. But when enthusing about the 20 to 30 per cent of landholders who are in Landcare we must reflect on the 70 to 80 per cent who are not.

The question is: how can land conservation be made comprehensive? Perhaps another way of asking the question is: should it be left optional? There can be little doubt that the community at large would take the view that whether land should be managed sustainably or degraded is not simply a matter of individual choice. Landcare falls a long way short of coming to terms with this issue and does not have significant prospect of making land conservation truly comprehensive.

Similarly, Landcare as it is structured may not ensure that the community comes to a proper understanding of what sustainable use of land really means. This is not something that can be rushed. It will take time and hard work. It will occur only if all involved have to address and go on addressing rigorous criteria to ensure that they think through the hard questions. Some of this may occur under Landcare, but not enough. The problem is summed up in the suggestion that Landcare is inclined to be all things to all people. In short Landcare tends to lack rigour.

Associated with these difficulties is the need to ensure that land conservation will be sustained. What happens when the money runs out? Some Landcare activities will no doubt continue. But systems like Landcare rely heavily on the energy and skill of relatively small numbers of people. A method is required to ensure that when they flag, the work goes on.

The goal must be not just a Decade of Landcare in which we can run a long hard race and finish. The decade must be seen not as a period in which to solve Australia's land conservation problems, but as one in which to put in place a comprehensive, rigorous, ongoing system which will ensure indefinite sustainable use. One must question whether Landcare is capable of achieving this goal.

Landcare and crisis

The greatest difficulty with Landcare may be not that it will be ineffective, but that it will be effective enough to prevent the establishment of a truly effective system of land conservation until the next crisis. Ironically, on this view, the greater the effectiveness over the Decade of Landcare the more likely this may be, and the greater the next crisis will need to be before really effective action is taken.

Land conservation is not, like much else environmental, a creature of the last decade or two. It goes back thousands of years, and in modern times it goes back most recently to the legislation and programmes developed since the 1930s and 40s. There is a history to learn from. What we seem to see is variations on a theme. A crisis is followed by concern, good intentions and some action. There are some gains, concern is eased, and action fades until the next crisis.

Landcare is now very popular, but the degree of awareness and concern with land degradation in the last major crisis of the 1930s and 40s was enormous. There were demonstration farms and soil conservation prizes, impressive rhetoric, and fine policies. There were active local groups and heroic performances by individuals. There was extension and some effective action. But because an effective, comprehensive, ongoing system of soil conservation was not put in place, the action slowly subsided and short-term exploitation took over—until the present crisis.

Some people insist that the problems of the past were different, that we overcame them and that now we are dealing with new problems. That is at best a half truth, at worst dishonest. Take just one example, salinity. It is now a huge problem, not least in Western Australia. Much of the problem has developed as a result of the massive land clearance that has taken place since World War II. Yet the Premier of Western Australia in Parliament in 1945 gave a concise definition of the cause of dryland salinity which could not be bettered.[24] As unbridled clearance continued, he pointed out that the Department of Agriculture was concerned and dealing with the problem.

The essential point is this. Knowledge and concern have existed before. They have achieved something, enough to stave off the establishment of a comprehensive, rigorous, sustained land conservation system until the next crisis. No one can be dogmatic, but we must wonder whether, with Landcare, we are doing it again.

The Use of Law

In order to overcome the difficulties with Landcare and to bring about and to sustain comprehensive sustainable land management, it is necessary to use the law. To this end, the law must be reasonably flexible, but sufficiently rigorous. It must embrace catchment or district-wide holism but also attend to particular problems and individual properties. It must also ensure reasonable objectivity and accountability while allowing scope for individual initiative and responsibility. Above all, it must ensure that proper consideration of the meaning and effect of sustainability will continue indefinitely.

The closest that Australia comes to such a programme is found in the land management legislative package in South Australia consisting of the *Soil Conservation and Land Care Act* 1989, the *Pastoral Management and Conservation Act* 1989 and the *Native Vegetation Act* 1991. This package has its deficiencies, but it is heading along the right path. In broad terms these pieces of legislation operate as follows.

Land conservation

The *Pastoral Management and Conservation Act* operates only in the pastoral lands, those more arid areas used almost entirely for low-level grazing. It sets up the leasehold tenure system and, more particularly, requires the land to be assessed 'in accordance with recognized scientific principles' as to its capacity to withstand grazing pressure (Sec. 6). This is taken by the primary administrative body, the Pastoral Board, to include grazing by native and feral animals. The principles guiding the assessment process are set out in the objectives of the act and the definitions of sustainable use. The key feature of this act is the process of scientific assessment, which must be undertaken on a periodic basis (Sec. 25). It seeks to ensure that grazing capacity and sustainable use are being thought through with a rigour which has not occurred before. It is interesting to note that the main test for sustainability is biodiversity, the maintenance of indigenous plant and animal life (Sec. 4. b. ii).

In overall terms the *Soil Conservation and Land Care Act,* which applies throughout the state, including the pastoral zone, is a more significant piece of legislation. It makes provision for the whole state to be divided into districts which, where appropriate, can embrace catchments, and provides for the establishment of a seven-member Soil Conservation Board for each district (Sec. 22). Board members must consist largely of landholders residing in the district. Boards are given some government funding, though they are essentially voluntary. Technical support is provided by the Department of Agriculture.

Soil Boards are required under the act to prepare a district plan within five years (Sec. 36). The act specifies the criteria to be taken into account in performing this function. They include: the nature and extent of land degradation in the district; the classes of land within the district; the preferred uses of land, that is the uses to which the land can be put on a sustainable basis; and management practices required to ensure that land is used within its sustainable capacity. Boards are also required to prepare a three-year action plan. Both plans are required to be reviewed regularly (Sec. 36). The act also establishes a twelve-member Soil Conservation Council, independent of government direction, which represents a wide cross-section of the community. Among its supervisory functions it must comment on district plans before their adoption.

The act makes provision for encouraging landholders to prepare property plans, and empowers boards to require such plans (Sec. 37, 38). Boards are also empowered to issue soil conservation orders. Such orders can be issued by a Soil Conservator if boards fail to act (Sec. 40). The act also places a general legal duty on landholders to use land with reasonable care to prevent degradation.

The most significant element of this act is the district planning system. Its greatest strength lies in the statutory criteria that must be addressed, and in the requirement that they must go on being addressed. The criteria give the process its rigour. The effect is that the truly long-term question of sustainability must be grappled with. Also crucial is the process of regular review which recognizes the need to ensure the steady ongoing accumulation of wisdom and action. It is one of the great failings in the past that energy flags or satisfaction sets in. This review process should ensure that the hard questions continue to be rigorously thought through and acted upon.

However, there are several weaknesses in the act. The first concerns property planning, which is crucial to sustainable management. Unfortunately the act structures property planning as if it were a form of punishment, and may well discourage it. It should abandon the concepts of compulsory and voluntary property planning and their present complexity, and provide simply that, just as all districts have a district plan, so all properties have a property plan. It must become accepted that property planning is not a penalty but an essential aspect of sustainable management. The primary benefit from such plans would, it is suggested, flow not from their existence, but from the need to address specific criteria and the process of review on a regular basis. They would, of course, also require all landholders to come to terms with and take into account the district plan.

A second weakness in the act concerns the translation of accumulating knowledge and wisdom into on-ground action. There is no obligation to give effect to a district plan, no matter how much it emphasizes the significance of a particular issue or practice. Both in managing land and in preparing property plans, landholders should be required to 'take into account' the district plan. This concept is not too difficult; it allows wide scope for flexibility and individual initiative, but is consistent with district-wide sustainability. Landholders should be expected to 'take into account' their property plan in managing their property. Again this would not require rigid adherence, but rather a genuine attempt to give effect to the plan. It should not be objected that plans will not be given any rigour if they merely have to be taken into account. Their greatest benefit may well be in thinking through the issues in their preparation. But it is not unreasonable, given the importance of land, to expect some level of plan compliance.

A third weakness concerns soil conservation orders, which may be issued by boards as a means of enforcement. The power to issue orders is extremely wide. Such orders have existed in various jurisdictions in Australia for fifty years. History shows that, as a means by which to achieve better land management, they have never been effective. It strongly suggests that they will be used only for problems approaching disaster status. Thus there is a huge gulf between power and practice. Soil orders should be used to enforce, rather than make law. This may involve greater clarification of landholders' obligations.[25]

It is necessary, too, to abandon this last-resort mentality of land conservation legislation. The aim should be not to throw out the odd bad apple, but to ensure the overall steady improvement of land management, with greater emphasis on property plans. Conservation orders will continue to be available, but the emphasis should not be on their issue but rather on their role in ensuring serious discussion and negotiation at an earlier stage. Thus, for board members and other key players, discussion and negotiation skills should rank highly.

Biodiversity conservation

Pastoral assessments and district plans may take into account the whole range of land management issues, including biodiversity aspects of native vegetation, and they will no doubt increasingly do so. However, the conservation of biodiversity through native vegetation protection is a complex question requiring specialist input, which occurs through the *Native Vegetation Act*.

This act is not just about keeping native vegetation for its own sake. It precludes the clearance of native vegetation only where it meets certain criteria. These include landscape or amenity value and significance in the conservation of land or water. But predominantly the criteria deal with the significance of vegetation for the conservation of biodiversity. Thus vegetation cannot be cleared if it contains high diversity of plant species, contains threatened plants or plant associations, provides significant habitat for wildlife, is associated with a wetland, and so on.[26]

Experience makes clear that biodiversity conservation will not be achieved if left to landholders or local decision-making. When a landholder applies to clear native vegetation, scientific officers from the Native Vegetation Management Branch in the Department of Environment assess the vegetation for its biological value and prepare a detailed report. The report goes to the applicant and to the Native Vegetation Council, which is responsible for making the decision. Before it makes its decision the council is obliged to hear representations from landholders and consults various bodies as appropriate. In each case it consults local government.

Landholders, even when presented with a strong biodiversity argument against clearance, still frequently express a wish to clear.[27] Even more striking is the position of local government. They will acknowledge that they do not have the expertise to carry out assessment. Yet when local government bodies which are clearly aware of the terms of the legislation are presented with detailed assessments demonstrating high biological value, the majority still say that they support clearance.

Looking into the future, it is argued that sustainable management of land will increasingly be required to take into account, and will indeed turn on, the conservation of biodiversity. The evidence suggests that the conservation of biodiversity will not be achieved unless it is supported by the law. Despite the very strong community support for wildlife and nature conservation, people generally do not have a good grasp of biodiversity issues such as the measuring and significance of the various categories of conservation significance (e.g. threatened, rare or uncommon). Even when the biological information in support of the protection of a threatened area is clearly presented, farmers and local authorities regularly argue the case for clearance. Furthermore, the South Australian experience shows the importance of specialist legislation and specialist bodies to deal with these issues. It is likely that they will be required for some time to come.

16

How Government Policy Can Effect Environmental Change

EVAN WALKER

There is no doubt that, in issues such as land degradation, well-based policy commitments at state and federal levels can generate remarkably effective programmes. My own experience in this regard provides a good example—the attack mounted on the salinity problem in the decade from 1980 to 1990, first by the state of Victoria and then by two of the other three states involved with the Murray–Darling Basin, and the federal government.

Because of my involvement in salinity control and in the Landcare programme which followed in Victoria, I will concentrate on the development of those two programmes. Of course, it would be equally appropriate to detail the approach taken to the problems of high country grazing, timber industry management, misuse of agricultural chemicals, or overstocking and overworking of marginal crop lands, because in each case the Victorian government developed a significant policy which gave rise to effective programmes. Unfortunately, not all of those examples have risen above the problems of adversarial politics. Thus, much good work is still subject to change with any change of government. But the work on the problem of salting is probably the best example to look at. Victoria did take a leading role in developing a concerted approach to salinity control and in creating a strong political consensus.

The Victorian Salinity Programme

My own personal awareness of the salinity problem began when I worked on wheat farms out of Swan Hill as a young man in the early 1950s. One of my farmer contacts then was very aware of the problems of poor irrigation practices. Even at that time there was opposition to the prospect of channelling saline run-off to Lake Tyrrell or lesser evaporative basins.

The event that remains in my memory as a clear indicator of the emotional nature of the issue was my attendance at a meeting of farmers in Kerang in 1980 when I was Victorian Opposition spokesperson on conservation. Farmers from that area were vehemently opposing the construction of the Mineral Reserves Evaporative Basin (MRB) which had been proposed by the State Rivers and Water Supply Commission and was supported by the all-party Public Works Committee of the Victorian parliament. Under local pressure that day the whole committee changed its view. I thought to myself, 'This is not the way to make long-term policy'. I began to get a real awareness of the immensity of the problem and the need to develop good policy and a solid process for tackling the problem at all levels. After a proper process had been developed and documented in the report *Salt of the Earth*,[1] tabled in cabinet, and good long-term policy had been prepared and agreed, the MRB proposal was halted and recast with agreement all round.

Before considering the salinity programme in detail, it is important to briefly tell the salinity story. Parts of Victoria have a geological history of naturally occurring high salt levels. The whole of the northwest of Victoria was once an ocean sea bed and there has always been some natural salt scalding. But as a result of massive clearing, salinization of land and watercourses has spread alarmingly. Water tables have been rising rapidly, and salinity problems are now being experienced in almost every region of the state, including Gippsland and the Western District.

It soon became clear to me that the problem was critical. If we did nothing, the further spread of salting presented a grim picture. Already three-quarters of the Kerang region and a quarter of the Shepparton region were affected. By the time the Labor Party won government in 1982, the direct cost of salinity damage to Victoria's agricultural economy was more than $50 million per year and was cumulative. Indirect costs were much greater, and unless the problem was successfully tackled these costs would treble every thirty years. Even worse would be the loss of land for future Victorians, the degradation of the environment, loss of vegetation and loss of wildlife habitat—to say nothing of salty drinking water.

It is not surprising that some remarkable solutions had been suggested over the years. One unlikely (but not quite hare-brained) proposal was to build an immensely expensive pipeline from the worst-affected area (at a cost of some $800 million) to carry saline water to the sea. This engineering solution was, to say the least, very expensive and not very elegant. Another proposal (which does fit the 'hare-brained' label) was to dig a second Murray River, and have one clean and one 'dirty'.

The problem has a long history in other parts of the world. Were you to visit Iraq today, you would find that much of their once fertile farmland is now abandoned. Yet it was on the plain of the Tigris and Euphrates Rivers (a region of the world then called Mesopotamia) that people first learned to raise abundant crops in dry land by diverting river water onto their fields. That was six or seven thousand years ago, and because of irrigation the region's fertility was legendary. Who has not heard of the Hanging Gardens of Babylon, the great cities of Persepolis and Nineveh, the mighty empires of Xerxes and Darius? At its height, Mesopotamia supported twenty-five million people when the world's population was a mere fraction of today's teeming billions.

Of course not all long-term irrigation projects have resulted in failure and the creation of deserts. For instance, Pakistan, whose River Indus plain is by far the largest continuously irrigated region on earth, is also a dry land dependent on di-version of river waters for the feeding of vast populations. Indications are that, if care is taken, irrigation can continue indefinitely. The more usual story is, regrettably, that where people have introduced large irrigation schemes onto vast areas of dry land, or have over-cleared upper catchments, salinity problems develop, reducing production and leading to the loss of vast areas of once good land.

Let me remind you of a more recent and huge tragedy of this kind. Some of you will have read the shocking story of the depletion of the Aral Sea in the former USSR, and the ruination of huge areas of once-productive land due to salting and misuse of agricultural chemicals. In the last twenty years that inland sea, once the fourth-largest in the world, has reduced in size by 60 per cent, and even if the crucial management and financial decisions critically needed were taken immediately, only a fraction of the former area could be regained. For the most part, the problems are immense and irreversible.

So the lessons are obvious. Salinity must not be allowed to take hold. Having understood that, and knowing that you cannot expect overnight success, the way the Victorian government tackled salinity is instructive.

The previous Victorian government had shown some concern and had referred the matter of the Mineral Reserves Evaporation Basin near Kerang to the parliamentary Public Works Committee. This helped to raise consciousness of the problems. But as we approached government in 1982, I made it clear (in my role as Shadow Minister for Conservation) that tackling salinity would be number-one priority. The chronology of events which led to the development of a comprehensive salinity programme is as follows.

1. In 1982 the new Labor government established an all-party parliamentary committee of enquiry into salinity with wide terms of reference. Over two and a half years it produced the report *Salt of the Earth* which described the situation very carefully and offered sensible recommendations for action.[2]

2. In 1985, following re-election, the government established a ministerial task force on salinity (not a statutory authority as recommended, since we wanted direct cabinet action). I chaired that task force and it included Ministers Joan Kirner (Conservation, Forests and Lands) and Andrew McCutcheon (Water Resources).

3. At the same time, the government established the salinity unit in the Premier's Department, and went to the key community leaders in affected areas to elicit their involvement. We understood the need for public education. We

convinced cabinet to break tradition and institute a co-ordinated salinity budget, overcoming, for the first time, the territoriality of portfolio budgets. The co-ordinated budget was made up of direct transfers from six portfolios.

4. We began a pilot programme in the Shepparton region, which was managed locally by a selected group of farmers, local people, and local government bodies, and was serviced by professionals from government agencies.

5. We convinced our neighbouring states to join us in calling a conference in November 1985 in Adelaide (which included the federal government) to plan a joint attack. From that conference, the Murray–Darling Basin Ministerial Council was established, with three ministers from each state and from the Commonwealth. It was useful that all four governments happened to be Labor. This did offer a political 'window of opportunity' that does not often happen.

6. We stressed the need for communities who were directly affected to 'own' the problems and process. State and national community consultation processes were begun. Key environment bodies were involved.

7. By late 1986 (in Victoria) we upgraded the ministerial task force to become a standing committee of cabinet—the Natural Resources and Environment Committee.

8. We developed in Victoria a catchment approach so that cause and effect could be handled together. The state was divided into nine regions in order to link local into regional strategies.

9. By January 1988, the Murray–Darling Basin Commission had been established to replace the River Murray Commission, and a huge amount of work was under way to develop a land- and water-based strategy for the whole Murray–Darling Basin.

10. By mid-1988, Victoria had published its state strategy, *Salt Action—Joint Action*,[3] to tackle the problem, and the Murray–Darling Basin Ministerial Council had drafted its strategy.

11. In all states and federally, budgets had been doubled and trebled to tackle the problem—and communities and individuals contributed their part.

12. Across the Murray–Darling Basin work was under way on farms with the assistance of farmers. Communities worked hard to develop management plans. States and regions even agreed to fund downstream mitigation works and agreed on salt disposal budgets.

The story, of course, can be told here only in outline. Suffice to say that, if all goes well, within ten to twelve years the spread of salinity will have been reduced to a minimum—and in some cases we will have begun to turn the corner and reclaim some of the lost land. The stakes are high when you realize that:

- the Murray–Darling Basin comprises one-seventh of the surface of Australia;
- it produces nearly 40 per cent of the nation's primary produce;
- the value of its production annually is some $10 000 million;
- losses each year already total over $250 million;

I have deliberately not discussed technical details. However, the whole of the basin is now being managed to conserve resources of both land and water. Co-ordination of management exists from the individual farmer to the community, to the region, to the state, to the nation.[4]

This experience is an object lesson in tackling a seemingly impossible problem. If I am right, we will still be feeding the nation and exporting from areas of land that would otherwise have rapidly become desert. The lessons learned are many, and it is important to realize that years of commitment by governments of all kinds lie ahead. The whole community must be made aware of the problem and how it is being tackled. To come as far as we have, a good policy-making process and a sound policy base have been essential.

Landcare

The work on salinity gave rise in Victoria to the Landcare approach. Care of the land had always been a priority in Victoria, but with the integration of the Soil Conservation Authority and the Vermin and Noxious Weeds Board into the Conservation, Forests and Lands portfolio, and with the growth of local action groups of all kinds, including the salinity groups, there was the opportunity to develop a more comprehensive approach. In 1985 Minister Kirner took the initiative to build the Landcare programme.

From memory, the first group to formally take the Landcare name was at St Arnaud. Since then some 350 community-based groups have been formed, working on their own land (and related private and public land) to tackle priority problems such as weeds, ragwort, rabbits. Support for this approach is widespread. The Victorian Farmers' Federation was involved from the beginning, as was the Australian Conservation Foundation. But it is important to say that action followed development of government policy and government initiation.

Assistance offered to groups (which might include an average of twenty farmers in an area) includes technical information, expert advice, assistance grants and help with approaching other funding sources. It was decided to use the nine salinity regions to co-ordinate the work around the state. The Victorian Landcare initiative was so successful it was taken up by all other states.

The setting up of Landcare was followed in 1987 by the publication of the State Conservation Strategy document, *Protecting the Environment.*[5] This has since formed the policy base for long-term rural and urban initiatives to ensure that Victorians learn to use resources wisely. The *Victorian Decade of Landcare Plan* was released in 1992. It is a logical progression from the earlier *Salt Action—Joint Action,* and provides opportunities for local and regional communities and government to co-operate in identifying priorities and developing action programmes. The success of the strategy is to be evaluated in 1994, 1997 and 2000. It is expected that succeeding governments, of whatever colour, will maintain the programme.

Like the salinity programme, the Landcare programme has been extended to become national. At the July 1990 meeting of the Australian Soil Conservation Council, ministers from all states and the Commonwealth agreed to the preparation of a national *Decade of Landcare Plan.*[6] Each state has produced its own plan under common guidelines, and these plans, in combination with the Commonwealth overview, form the national plan.

In Victoria we now have the State Landcare Committee, made up exclusively of community representatives and advised by government experts. Represented are the Municipal Association of Victoria, the Victorian Farmers' Federation, the Australian Conservation Foundation, the Conservation Council of Victoria, and both the Dryland and Irrigated Land Salinity Committees. It is hoped that eventually this body will become the umbrella body for all related groups in the state. There is also a standing committee of all key government agencies which is convened by the Department of Premier and Cabinet. It reports through the cabinet's Natural Resources and Environment Committee direct to cabinet. The work of the standing committee and the cabinet committee is comprehensive and affects both public and private land.

Of course the story is not perfect (nor is it all told yet). There are problems. Apart from the normal funding difficulties, some recent policy actions have been of doubtful value. It is clear that government policy can affect moves for environmental change negatively as well as positively. Some features of the recent water legislation are too narrow and commercial, and reflect the new 'economic rationalist' approach. Indeed one might say that the mandate and management of the Rural Water Commission itself is far too narrow. The new act does not seem to tackle problems of a broad kind.

Nevertheless, the story overall in Victoria is good, and reflects the value of well-based policy-making in the business of long-term good management of the land. It is possible to say that most of what has been accomplished in Victoria in the last decade could not have happened under conservative governments. The political realities are simple. Tough political decisions are almost impossible to take when seats are at stake, and it was apparent to me (and others) at the beginning of the 1980s that we might be enjoying a rare, if not unique, political opportunity with three state Labor governments and a Labor federal government. It remains to be seen whether significant change to those governments will undermine what must be seen as long-term programmes. I hope not. The fact that New South Wales continued with the national salinity programmes after the change of government in 1988 was heartening.

Conclusions

How can government policy can most effectively bring about environmental change? My response to that question is to say that to effect change, policy must have three characteristics:

1. It must be well-based, intelligent policy.
2. It must be implemented enthusiastically, using good process at all levels.
3. It must be long-term—and therefore supported openly by all major parties.

This has been the basis of success in both the salinity and the Landcare programmes. It will take another decade to be sure of their long-term effectiveness.

17

Changing Cultures in the Farming Community

SHARMAN STONE

I grew up on the family farm on the northern Victorian Tragowel Plains in the 1950s. In those days the community attributed 'the salt' to poor farm management. Since the state of the farm related directly to the family's status and standing in the community, a patch of salty ground in the back paddock was a best-kept secret.

In the 1970s I came to study the effect that environmental conditions had on the organization of some of these northern Victorian communities, on local government leadership and on individual farming families.[1] At that time the traditional farmer values of independence of action, or 'being your own boss', continued to be the most highly valued of the attributes associated with the farming way of life. At the same time landholders were deeply stressed by their inability to solve their 'salt problem', and were frustrated by their inability to control the irrigation practices of some neighbours who regularly 'flooded them out'.

In the 1980s I worked with the state salinity programme, developing community involvement strategies, and again researching the social consequences of soil salinization in rural communities. I was there to witness the revolution. In 1988 the Tragowel Plains community proudly celebrated the centenary of their public irrigation system. At the same time the community publicly acknowledged the devastating consequences of the high saline water table, in large part a legacy of that pioneering irrigation scheme, and together they pledged to implement their locally developed Salinity Action Plan.

Shared responsibility and co-operative action by land-holders to solve a problem like soil salinization requires nothing short of a revolution in rural community thinking and behaviour. Achieving co-operation between government departments supporting such community action also requires a fundamentally new approach to environmental management. In the mid-1980s the Victorian government implemented the state salinity programme. The Tragowel Plains community was given some resources and technical assistance to produce a multi-faceted environmental plan. As the landholders participated in the planning process, their traditional independence and the stigma associated with 'having a salty farm' gave way to a co-operative approach to farming and drainage control, and pride in their significant achievement as pioneers in managing a saline agricultural environment.

The revolution was under way.

Traditional Rural Values

The farming ethos

Thirty years ago farming in Australia was substantially different from other occupations. Farmers had developed codes of conduct to help one another survive short-lived disasters such as injury, flood or fire. Their typical response to catastrophe helped to establish the bushman legends which attributed self-reliance, independence, egalitarianism, mateship, conservatism and resourcefulness to all Australian people. Farmers came to attribute substantial non-monetary benefits to farming as a way of life. To survive emotionally and financially, they developed long-term planning horizons, being prepared to work and invest for the greater security of the next generation. Handing on the family farm was a moral obligation and the point of great striving. 'Losing' the family farm was shameful and a horror almost beyond contemplation. Farmers' children had different school retention rates and aspirations. Rural people's response to serious injury or ill-health differed to that of urban populations, and they could be distinguished by the way they dressed, their speech and manner.

Thirty years ago the farm/non-farm, urban/rural delineation in society was seen as logical, whether based on technological, economic, political or social grounds. Our poets and playwrights celebrated the differences with tongue-in-cheek humour, reverence or riotous send-up. City people learned about the farmer's unique contribution and country culture through visits to country cousins and a regular dose of *Blue Hills* or *Dad and Dave* on the wireless. (These differences are still often caricatured in national advertising and television drama.) Farmers had fewer conveniences than town and city dwellers in the early days, but these were not so much the subject of invidious comparisons; rather they were accepted as the consequences and attributes of a special way of life.

Australian farmers considered themselves, and were considered by others, as 'uniquely worthy'. They reasoned that they worked hard to produce 'essential' food and fibre for a growing nation. By the sweat of their brow and ingenuity they tamed the bush and turned the deserts gold with grain. The 'man on the land' battled flood, fire and plague, not only to feed and clothe the country but also to create national wealth through growing vital exports. And they were enormously effective: forty years ago 200 000 farms produced 25 per cent of Australia's gross national product and 95 per cent of all export earnings. The country rode high on the sheep's back, and everyone was grateful. The belief in farmer 'uniqueness' manifested itself, in particular, in the political process. Until the 1970s farmer spokesmen held the political initiative and controlled the farm policy agenda. Dedicated institutions evolved to serve the special agriculture and rural community needs: Departments of Agriculture, Colleges of Agriculture and Horticulture, the Country Roads Boards, Rural Water Commissions, Soil Conservation Authorities, Country Fire Authorities, Bush Nursing Hospitals, Consolidated Schools, and a Rural Finance Corporation.

An agricultural creed that supports farmers' claims to a unique and separate status within Australian society retains currency. This creed is not exclusive to Australia, but has been evoked by and ascribed to farmers in most developed countries, for example in the USA, New Zealand and Canada.[2] The main tenets of this creed are:

- farming life is a healthy, 'natural' human existence;
- farm life is superior to life in a city;
- farming is not only a business but a way of life;
- farming ideally is an intergenerational family enterprise;
- it is good to 'make two blades of grass where only one grew before';
- anyone who wants to farm should be free to do so, without interference;
- a good farmer is self-reliant, independent and his own boss.

While its special ideological status makes the creed useful in selling anything from socks to breakfast cereals, it does little to help forge the strategic alliances with consumers, opinion leaders and policy-makers which would help Australians to understand rural problems and share responsibility for future action.[3] At a more practical level, alliances could help recruit workers to assist in environmental rehabilitation works and provide the public support needed for expenditure on regional planning and conservation work.

Prior to the 1960s there was little public discussion about food and fibre production or their environmental consequences in Australia. Classical economics taught that producers responded automatically to the needs of the consumer through the mechanism of the market place. It was then assumed that if the country saw to the needs of the producers, consumer well-being would automatically follow. Given that farmers made no secret of their intention to pass on their family's land to their sons, it was also presumed that none would knowingly degrade their children's legacy by environmentally unsound practices. Despite the fact that land degradation caused by high saline water tables and soil erosion had been evident since the turn of the century in some areas such as northern Victoria, consumers did not concern themselves about the sustainability of agricultural practices, and advocates did not emerge to protect their interests.

Twenty years ago farmers began to lose control of the agenda for agriculture and natural resources. Advocates emerged to promote environment, alternative life-styles, animal liberation, health and fitness, organic methods, occupational health and safety, trade unions, agribusiness and value-adding. Each vied for the attention of the media and the government, seeking to influence research priorities and the expenditure of the ever-diminishing budgets of Departments of Agriculture. Farmers typically responded to these lobby groups and advocates as implacable enemies, bent on driving the last vestiges of free enterprise and self-determination out of the hands of the 'man on the land'. Even where farmers had been making concerted conservation efforts for years and had proof of their ability to protect remnant vegetation or rehabilitate it, they felt politically and philosophically alien to 'the greenies'. They could not embrace them as 'brothers and sisters' in the struggle for a sustainable ecology. The success of the various advocates in shaping public opinion in the 1970s and 1980s deepened the farmers' sense of being misunderstood and marginalized. Their sense of powerlessness grew.

Today there is a growing gulf between the reality and the rhetoric that suggests Australian farming is a unique and satisfying way of life. Primary producers are experiencing low returns, rising input costs, growing farm indebtedness, high interest rates and declining land values. With the long-term future also clouded by the consequences of some unsustainable agricultural practices and by insecure export markets, it is little wonder that meetings of farmers across Australia frequently give voice to feelings of despair, anger and bewilderment. They cannot understand society's apparent failure to protect their interests and the indifference to their plight.

No-one needs to be reminded that times are very tough in the bush today, and country towns with fewer than ten thousand people are losing population and services. Families are leaving agriculture at unprecedented rates, financially crippled by free-falling incomes, high land values, and punitive interest rates on the 'get-big-or-get out' loans offered in better times. It is hardly surprising that most farm children are educated with a view to equipping them for life off the farm. Farming is now recognized as one of the most stressful and hazardous of occupations in Australia, with the incidence of serious accidents, allergies and mental health problems greater than in the general population.

The physical isolation, self-sufficiency and differences in consumer choices in the country are now largely diminished. With improved communications and transport, and chain stores predominating, farming populations eat the same food, buy the same clothes, play the same sports, study the same school curriculum, watch the same television, read the same press, and want the same amenities as urban populations. However, farmers' ability to match the standards of living of urban populations is declining. Weeds, the weather, insects, disease and the fluctuations in markets manipulated by foreign governments make most farming a wholly unpredictable enterprise. In the non-farm sector, economic risk for individuals has been reduced, with job contracts, unemployment payments, minimum wages and working conditions. It is difficult for a rising generation of farmers to accept the financial risks of primary production while the rest of the economy is shielded from absolute penury. For the younger, better-educated, more innovative, and commercially orientated farmers, the traditional beliefs in the value of independence and the superiority of farm life no longer provide enough psychic income to compensate for its lower dollar returns, higher risk, and lesser convenience.

Changes in rural community organization and social structure

In fully productive areas, and in better days, rural adjustment took place as smaller farms grew bigger, or incoming families were slowly integrated through their involvement in a plethora of community sporting, service, social, church and other voluntary organizations. Such activity provided recreation, reduced social isolation, and offered a regular opportunity for discussing farm issues with neighbours.

My studies of the salt-affected areas of northern Victoria showed that, while traditional values remained, deteriorating land and water resources triggered a particular pattern of selective depopulation in response to the less favourable long-term prospects of the area. While skilled and experienced farmers tended to replace exiting farmers in the days of good returns and prospects, newcomers buying into areas of environmentally degraded land tended to be less experienced and undercapitalized. As the population fell below the numbers necessary to keep the schools, shops, and community organizations viable, the opportunities for these newcomers to interact with and learn from more successful and experienced farmers dwindled.

The younger, less-experienced, debt-burdened farmers quickly sought a new start elsewhere. If they failed to sell, they sought off-farm work, leaving intensive irrigation and farm maintenance undone.

The more innovative, long-established farmers who invested in laser grading, irrigation water re-use systems, turkey-nest dams, or new crop types tended to become overwhelmed by the growing burden of interest due on debt, and under-whelmed by improved productivity. They also sought to sell. The loss of these role models in the area had serious implications for the diffusion of new technology.

The same form of adjustment took place in the salt-affected towns, where the more astute business people assessed their long-term prospects and sold to others with less acumen or reduced expectations. Businesses fell vacant. The government departments responded to the falling population by 'region-alizing' service to a bigger centre, or replacing it with a part-time or mobile facility. The public sector employment opportunities and ready access were lost to the towns.

The younger generation of local farming men and women joined the newcomers in seeking income off the farms, depending on their parents to run the family property into advanced old age. Neither the old nor the young could continue to dedi-cate the time necessary to maintain the community infra-structure and the network of clubs and societies which had once welcomed newcomers and provided a forum for local infor-mation exchange. Communities need support and information for their community leadership, extra 'manpower', and some financial resources to tackle the environmental problems. Without this they must continue in their belt-tightening, con-servative-thinking, survival mode.

Farmers working to the limits of their physical and financial resources cannot achieve the fencing of remnant vegetation, reafforestation, stream-bank stabilization, soil and pasture reno-vation, pest and weed control and other works which they often realize are necessary if the country is to achieve its environmen-tal objectives. As they look over their fences into the state-controlled forests and national parks, landholders also observe that government agencies seem similarly unable to undertake the work required to eradicate vermin and noxious weeds, or to control soil erosion.

As an understanding of the underlying causes of land degradation becomes widespread, farmers also come to realize that their individual actions can do little to arrest rising regional water tables, native forest dieback or the degradation of river systems. Given this growing understanding and their contracting financial and physical resources, it is hardly surprising that many farming families choose not to invest in new environmentally sustainable technology which may also require them to forgo income while they reorganize their enterprise to become 'ecologically sustainable'.

The protection and rehabilitation of Australia's natural environment requires expert knowledge, co-ordination and co-operation between landholders, the managers of public lands, local government, and the forestry, mining and recreation industries. However, achieving this will require more than the time-honoured consultation strategies which rarely do more than preach to the converted. If environmental protection and rehabilitation is to be achieved, the public needs to support long-term research, planning and farmer assistance programmes with substantial levels of public investment.

Achieving Landholder Co-operation and Co-ordinated Action

The salinity problems in northern Victoria were first minuted in Kerang Shire Council proceedings at the turn of the century. (They complained that a neighbouring shire was the cause of the problem.) In the years that followed, hundreds of scientific reports charted and analysed the spreading impact of the high water table as it manifested itself in dying forests, crops and pastures. Detailed treatises mapped the underground aquifers and tracked their major source to the deforested Central Highlands hundreds of kilometres to the south. Year after year the northern Victorian communities saw drilling rigs from competing government departments line up to monitor the steady saline flows from bores beside their boggy country lanes.

It was not until the state salinity programme of the mid-1980s that the salt-affected farmers were given adequate resources and asked to participate in developing a plan to tackle their local salinity problem. The community-led planning process empowered the community. It reconnected farming families and individuals as it gave them collective responsibility for advice and actions in their local area. It gave them the information and the confidence to challenge and replace the traditional farmer response with a co-operative and supportive approach which swept away the stigma of farming in an environmentally degraded area. Salinity came out of the closet as farmers in neighbourhood clusters pored over aerial photos and invited each other to walk their property, to learn from their own lifetime of trial and error, struggle and experience.

The engagement of landholders in salinity planning required consultation strategies which were significantly different from the approaches commonly taken by agricultural extension services or movements like Landcare. The Victorian salinity programme recognized that waiting for people to respond to invitations, or 'targeting high flyers' would not change the rural community's highly selective participation rates. Landcare appealed to the individual who was already motivated and had the confidence and the time to participate. It typically preached to the converted, especially in its establishment stages, which could last for the first two to three years of the group's life.

The salinity programme had to involve all landholders, including those whose management practices needed to change. These were often the people with the least ability to invest time or resources into new farm works. Such people were often aware of the need for change, but preferred not to expose their personal circumstances to the local Landcare or Agricultural Department discussion group. There was no 'loss of face' in their staying away. A 'traditional' farmer can readily demean a discussion group or the local Landcare membership, labelling them 'over-zealous', 'greenie hobby-farmers', or 'mostly women'.

The state salinity programme had to develop new technical information. It had to tap and disseminate the special knowledge gained by generations farming in particular environments.

It had to engender a spirit of co-operation and bring about genuine participation in environmental planning amongst landholders, interest groups and bureaucrats much more accustomed to competition and independent action.

To achieve community-led planning and a grassroots commitment to action, the state salinity strategy established community Working Groups of twelve to twenty-five elected representatives of the key stakeholder groups in each salt-affected area. The Working Groups were specially resourced over two to three years to produce a detailed plan for salinity mitigation in their sub-region, subject to agreed down-basin and cost recovery considerations. A technical support group of departmental specialists supported each working group, testing their proposals and options, and assembling data necessary for them to have a workable final strategy. Sub-regions ranged from four hundred landholders to many thousands, and included the Kerang Lakes, the Barr Creek, the Tragowel Plains, Nangiloc–Colignan, Sunraysia, Campaspe West, and the Goulburn Irrigation and Dryland Regions.

The key to community participation was the requirement that a Working Group's salinity strategy had to have community endorsement before it could be submitted to state cabinet for funding. The first Salinity Working Group, in the Barr Creek, sought to obtain 'community endorsement' through the time-honoured method of exposing their final draft plan to local public meetings. Their efforts were met with the usual apathy and traditional resentment reserved for those who advertised their area's environmental problems.

This initial failure led the Working Group to try a totally different approach. Using so-called Local Action and Advisory Groups (LAAGs) we achieved landholder participation rates of over 95 per cent, and such speedy implementation of the strategy that targets were exceeded. LAAGs then became the means of involving landholders in planning strategies for both salinity and floods throughout northern Victoria.

LAAGs: What did they achieve?

LAAGs consist of clusters of twelve or so farm enterprises. All clusters abut to form a continuous patchwork over the sub-regional area. Consequently every sub-regional landholder, big or small, part or full-time, is a member of a LAAG. The boundaries between LAAGs are usually delineated with the assistance of the Working Group, who also identify a potential leader from each of the LAAGs. The nominee is then asked to take up this voluntary role of co-ordinator and official spokesperson for the LAAG. In my experience of approaching more than three hundred potential LAAG leaders, only two declined the offer, both with sincere regret.

LAAG leaders are given a detailed briefing by the Working Group or lead agency, and receive specially prepared written information for their membership, along with photographs, maps and whatever else is necessary to inform the group fully of the issues, the planning task and the options under consideration. It is the leader's role to ensure that each member of the LAAG is fully briefed, and every individual's response to the Working Group's questions or proposals is fed back within an agreed period, usually three to four weeks. Leaders are asked to contact every member of their LAAG *in person or by telephone* when inviting them to the LAAG meeting, held at a time and place of their choosing. Often the first meeting is around a barbecue, with much of the time spent re-establishing or initiating acquaintanceships. Some LAAGs meet a number of times during the period. They may invite members of the Working Group or other specialists to help to explain the issues.

The personal invitation to participate with peers in discussing issues of significant local consequence is very compelling. Usually at least one representative of every LAAG farm enterprise attends, but often whole families participate. Individuals unable to attend send along their questions and comments, which are duly recorded by the leader. While the need to record the response of every individual rather than a group response is stipulated, many LAAGs choose to give a single consensus response which they then actively promote to their neighbouring LAAGs and the Working Group. At the end of the period the leaders meet to debrief the Working Group, registering the participation rates in their LAAG and the individual or collective feedback.

LAAGs succeed because they take traditional farmer values into account. They give participants equal status in advising and finally approving an area's environmental management strategy.

They provide an increasingly rare opportunity for social interaction with neighbours in a familiar and convenient setting.
Both men and women and the young and old can comfortably
attend. LAAGs appeal to the landholder's sense of place. They
rely on verbal interaction, the preferred and the most powerful
communication medium in human society, and they provide
visual and written information in a locally relevant and easily
understood form. This leads to a sense of personal involvement
and ownership of the plan, which facilitates information transfer
and rapid implementation of strategies.

Peer group pressure is focused in the LAAG as neighbours
pore over aerial photos of their farms, study water table depths
and tree losses, or take farm walks to examine work proposals
or problem areas. Individual farm practices come under consideration. Different land management outcomes become obvious. Through such interaction traditional farmers have come
to realize that part-time landholders share a commitment to the
local area, even though their life experience, values and politics
may not coincide. Many part-time farmers come to appreciate
more fully the responsibilities of land ownership in a farming
community.

The degree to which rural communities feel comfortable
with the LAAG process was demonstrated in the Barr Creek
catchment some time after the intensive rounds of meetings to
review and respond to the local salinity plan. Another very important issue confronted local farmers. They had to quickly
become familiar with the issues involved in transferable irrigation water rights, then register their response. A series of
state-wide one-day workshops had been chosen as the consultation and education strategy. One was planned for the Barr
Creek area. Nearly half of the thirty-two LAAG leaders took it
upon themselves to contact their memberships, reminding them
to attend. At the workshop they chose to participate in the small
group discussions as LAAG clusters, and were critical that the
LAAG process had not been utilized to involve the entire catchment farmer population. This workshop had the highest
attendance in the state, while drawing on one of the smallest
irrigation populations.

During the Avoca Flood Study, managed by the Rural Water
Commission, the majority of the LAAG leaders chose not to

leave the working group and the lead agency to its own deliber-
ations after the first round of LAAG consultation. Leaders, often
accompanied by some of their members, attended all later meetings as observers to ensure that their input was given due
consideration, and to keep their members fully informed of
planning progress.

LAAGs can also bring about more cost-effective action.
After considering the consultant's estimates for flood bank
removal in the Avoca Flood Study, LAAGs produced counter-
estimates; by using their own labour and equipment and
borrowing the shire graders, they did the work for considerably
less, and finished it sooner. Barr Creek LAAGs quickly moved
to considering group tendering for work to be done on individual whole farm plans.

Achieving interdepartmental co-operation and co-ordinated action

While the organization of LAAGs is comparatively simple and
readily achieved, the total revolution in community participation in complex environmental planning and rapid uptake
of mitigation measures comes at a price. The Victorian salinity
programme was the result of the commitment of some $22 mil-
lion per annum over the life of one parliament. However, this
budget was not just the pool of financial resources to pay for
specialist research and data collection. Its management and
allocation also became the means by which interdepartmental
co-operation and co-ordination was ensured.

An independent secretariat, the Salinity Bureau, was
established to co-ordinate the budget bids of the five participating government departments. These departments had to
fully disclose and co-ordinate their salinity research and mitigation programme bids as part of the funding application process.
Each budget bid had to pass through filters of interdepartmental and community committees, which assessed their merit
according to the statewide and sub-regional priorities of the
salinity programme. By the end of this process any work that
was non-essential, overlapping or replicated had been exposed,
along with the proposals which clearly met the criteria.

Conclusion

Some of the most enduring of traditional Australian rural values are a belief in their 'unique worth', their self-sufficiency, and their right to independent action. As well, farmers understood they were in competition with one another, each striving to supply a strictly limited market. These values and beliefs evolved in the days of farmers' greater geographic isolation and prosperity. Such values helped them to survive in a physical environment that brought flood, fire and plague in regular episodes, interspersed with rising or tumbling prices for their produce. Today, however, such values frequently inhibit the inter-farm co-operation which is necessary if environmentally sustainable agriculture is to replace the practices of earlier generations. This farmer co-operation is necessary, because environmental problems are rarely contained within the boundaries of one farm enterprise.

The Victorian salinity programme proved that there are strategies which can engage landholders in a way that brings them to co-operate and co-ordinate their on-farm land and water management. This required, however, a government commitment to the long-term resourcing of community-led environmental planning, research and some on-farm assistance.[4] Too often the financially stressed and increasingly marginalized Australian farmer is expected to undertake the massive task of private land rehabilitation alone. This is not a realistic expectation. When given the opportunity, urban populations have demonstrated their interest and willingness to help, but they too need education and co-ordination.

Australia needs a land ethic based upon the premise that the individual is a member of a community of interdependent parts, in turn dependent on a sustainable environment. Our system of ecology management should reflect that interdependence, and make it possible for all to interrelate and contribute. We no longer have the time to debate the exclusive rights of individuals, states, departments or interest groups.

There is an urgent need for all farmers to recognize that their interests converge with those of environmentalists and Australian society. At the same time the special economic and emotional relationship that farmers have with their land needs to be understood and taken into account. In order to communicate with and be embraced by the rest of Australian society, however, farmers may have to modify that part of their creed which proclaims their right to independent action, and their unique self-worth. Where this has been replaced by co-operation and joint action, the achievements have been monumental.

18

Conclusion: Environmental Values, Knowledge and Action

DAVID YENCKEN

In the first part of this book six authors present views about values and environmental sustainability. All differ in their emphases, but share a common view: that significant changes are needed to the value systems that have inspired Western attitudes to development and 'progress'. In the second part four different views are given of the types of knowledge required to achieve environmental sustainability. Scientific knowledge, it is argued in two of these four chapters, needs a sharper focus, a more rigorous application, and a broader, more interdisciplinary base. The two remaining chapters do not deny the importance of scientific knowledge but argue that it has limited usefulness in dealing with environmental problems. The way this knowledge is generated, valued and privileged is a social construct infused with values and assumptions which deserve challenge and review. Persistently, it is suggested, we are concentrating on the wrong questions. The third part of the book deals with specific questions of land degradation and sustainable agriculture.

This last chapter will examine a number of propositions drawn from arguments presented in the book and will explore to what degree they are supported by the evidence from the case study and other material. There are, however, some prior questions which have to be considered before it is possible to turn to the propositions.

As several of the authors show, we cannot begin with the assumption that there is agreement about the meaning of environment, or about the concept of sustainability, or about the nature and extent of land degradation, or about the meaning of environmental sustainability as applied to farming and pastoral lands, let alone agreement about the best means to achieve it. The disagreement is not just about the scientific evidence; it is much more deeply rooted. There are four questions in particular that have to be asked.

What do we mean by environment?

Concepts about nature and the external world surrounding human beings are fundamental to all cultures. In the West, Judeo-Christian and Greco-Roman traditions have combined to give us the intellectual and conceptual frameworks for our view of nature and our relationship to it. It has been persistently argued that both traditions have emphasized a separation between nature and humanity. Criticisms made of the Christian tradition have been that, in Christian theology:

- God—the locus of the holy or sacred—transcends nature.
- Nature is a profane artefact of a divine, craftsman-like creator. The essence of the natural world is informed matter: God divided and ordered an inert, plastic material—the void, waters, dust or clay.
- Man exclusively is created in the image of God and is, thus, segregated, essentially, from the rest of nature.
- Man is given dominion by God over nature.
- God commands man to subdue nature and multiply himself.
- The whole cognitive organization of the Judeo-Christian world view is political and hierarchical: God over man, man over nature—which results in a moral peck order or power structure.
- The image-of-God in man is the ground of man's *intrinsic* value. Since non-human natural entities lack the divine image, they are morally disenfranchised. They have, at best, instrumental value.
- The theologically based instrumentality of nature is compounded in the later Judeo-Christian tradition by Aristotelian–Thomistic teleology—rational life is the telos of nature and hence all the rest of nature exists as a means, a support system, for rational man.[1]

It has also been argued that the Christian tradition has placed its emphasis on belief rather than experience, and that this emphasis has reinforced the separation of humanity from nature.

> Early in its history Christianity rejected *gnosis* or direct experience of God for *pistis* or the trust of the will in certain revealed propositions about God. Spirit then, is distinguished from nature as the abstract from the concrete, and the things of the spirit are identified with the things of the mind—with the world of words and thought—symbols—which are seen, not as representing the concrete world, but underlying it.[2]

It has similarly been argued that the mechanistic world view of science espoused by Descartes, Newton and Laplace had its origin in atomistic images of nature in Greek thinking, and that the dualism in Christian thinking between spirit and nature and between humanity and the natural world has its counterpart in Greek dualism.[3] As an example of this dualism, Freya Mathews argues in Chapter 5 that 'In Greek philosophy human identity was divided into a "higher" mental component and a "lower" bodily component'. Humanity was distinguished from the rest of nature by its possession of reason or mind and was eventually seen as categorically distinct from and superior to the natural world.

These views of the separation of humanity and nature have now been challenged from many quarters, notably from the findings of twentieth-century science (from relativity theory, quantum mechanics, ecology and the Gaia theory). The Christian church, as Peter Hollingworth demonstrates in Chapter 4, is reviewing its traditional attitudes to nature and the environment, and is looking to other interpretations of the Christian tradition more supportive of holistic and ecologically based world views. Hollingworth and other theologians such as Steindl-Rast argue that it is not so much the Christian texts and sources that are at fault, but rather the interpretations given to them.[4] There are, similarly, other strands in Greek philosophy that are being appealed to for different interpretations of the relationship between humanity and nature.

As part of this process, the meaning of nature and environment has also been changing, from a definition of environment as the natural world separate from human beings to a definition

220   Restoring the Land

of environment as a natural system of which human beings form one small part only.

This transformation is, however, far from complete. The view of humanity as separate from nature is so deeply ingrained in Western thinking that acceptance of this change is at best partial, and even where it has been achieved intellectually it may not have been emotionally. Thus even in the definition and understanding of environment there are significant differences of views and values.

What do we mean by sustainability?

The concept 'sustainable development' was first given expression in the World Conservation Strategy prepared by the International Union for the Conservation of Nature and Natural Resources (IUCN), the United Nations Environment Program (UNEP) and the World Wildlife Fund (WWF) and launched in March 1980. Lee Talbot, former Director General of IUCN, the agency primarily responsible for the drafting of the strategy, has described the way the concept of sustainable development was introduced into the strategy. Talbot described the first draft of the strategy as a

> wildlife conservation textbook, for at the time many conservationists regarded development as the enemy to be opposed and many developers for their part regarded conservation as at best something to be ignored, or at worst an obstacle to progress. With each draft the two sides were brought closer and involved a process of education. The final draft represents a concensus [sic] between the practitioners of conservation and development and a concensus which would not have been possible without the educational experience which the development of the various drafts provided.[5]

Sustainable development is not a formulation agreeable to all conservationists. Many argue that sustainability should be a goal in its own right, unshackled to development: the notion of development is seen as one of the misplaced legacies of Western traditions which should be discarded as a necessary condition of a good and equitable society. There are examples of these views in this book, notably in William Lines' chapter. Sustainable development has also been criticized as a meaningless formula within which any number of totally antagonistic viewpoints can be accommodated. There is some force in both arguments.

The concept of sustainable development can, however, be looked at in another way. It can be seen as an inspired way in which a bridge might be built between two conflicting paradigms, between the paradigm that has underlain past Western approaches to the environment and an emerging new environmental paradigm. It could be seen as a way traditional core belief and values might be altered. A second important point, one which has been made many times from Third World countries, is that environmental objectives which exclude development, and which thus limit the potential of these countries to achieve material standards for their populations equivalent to the material standards of Western countries, are unacceptable.

Thus, interpretations of sustainability are not only value-laden but vary greatly. Despite this variation, the concept of sustainability, however defined, is now widely recognized. There is a slow but progressive acceptance of the Brundtland Commission's definition of sustainability: 'Sustainable development is development that meets the needs of the present without compromising the ability of future generations to meet their own needs.'[6]

Is there a problem of land degradation?

This book began with a statement about the problems of land degradation in Australia. To many this statement would seem unexceptionable. Respectable scientists have measured the extent of land degradation. Their findings have been verified many times by other scientists. However, even among scientists there is not total unanimity. Some agricultural scientists have, for example, argued that, if we compare soil productivity today with what it was two hundred years ago, it can be shown that farming practices have improved rather than degraded the quality of farm land. As Michael Webber observes in Chapter 11, 'what is degradation by one definition may not be degradation by another'. Economists may judge land degradation by yet different standards. Webber reminds us that 'Conventional opinion observes that land degradation is a problem, but many economists argue that it may not be an important problem . . .

that there is no evidence that rates of land degradation are not socially optimal'.

What does sustainability mean when applied to land degradation?

In this book there are many different views about sustainability as it is applied to land degradation. For Egan and Connor, sustainability involves the way energy, nutrients and agrichemicals are managed on farm land to avoid environmental problems such as pollution of surface and subsurface waters by agrichemicals, the contamination of products by agrichemicals, the misuse of fragile lands and salinization. For Bradsen the essence of agricultural sustainability is the protection of biodiversity on farm lands. Bradsen argues that biodiversity is best achieved by the retention of areas of native vegetation rather than by any new planting programme. While we might agree that both views need to be taken into account, they are widely divergent because each is looking at the problem from a quite different perspective. Thus the perspectives which different individuals bring to definitions of sustainable agricultural practice vary according to their backgrounds and professional training.

It will be evident that these are all contested interpretations infused with values. Values, core beliefs, and disciplinary or systemic perceptions thus influence not only our responses to environmental problems, but all the defining principles with which we approach them. According to the way we define those underlying concepts, so we are likely to assess the nature of environmental problems and to respond to them. Another way to approach these conflicts of opinion is to see them in the context of changing paradigmatic views of the world.

The Propositions

A number of important propositions can be derived from material presented in the book and from the literature on environmentally related behaviour and decision-making. These propositions can be seen as hypotheses whose validity might be tested from the evidence of the case study and other material in the book. The propositions are as follows.

1. There are different instrumental means of bringing about behavioural change, which include: attitudinal change through education and other processes; the use of economic incentives or deterrents; the use of science (knowledge); and the use of law and administrative process. Some changes may require the influence of one or two of these means only. Others may require all the means working in concert.
2. Those concerned with environmental change tend to perceive solutions as falling within their own expert province, and rarely grasp the importance of other means.
3. Environmental change requires a redefinition of knowledge which embraces local, practical and intuitive as well as expert knowledge.
4. Linear transfers of expert knowledge are not satisfactory means of achieving environmental change. Knowledge exchange needs to be seen as a process of mutual exchange between the users (here the farmers), the researchers and the intermediaries.
5. There is a need for greater status to be given to environmental research as an interdisciplinary activity.
6. Environmental problems are inherently social and political, not simply scientific. Thus environmental research should concentrate as much of its effort on the social and political as on the scientific issues, in the same way that environmental decision-making needs to be concerned with the social and economic, as much as with the scientific, aspects of environmental problems.
7. Sustained long-term behavioural change requires a change in core beliefs and values.
8. Dialogue is an essential part of changes in core beliefs and values.

Each of these propositions is discussed below.

The Different Means of Bringing about Behavioural Change

John Bradsen in Chapter 15 identifies a number of 'techniques of thought and social organization' which influence behavioural change. He nominates attitude change, science, economics and the law. After examining the limitations of attitudinal change, science, and economics, he discusses the potential of the law. He points to the many examples of attitudinal change which have not been accompanied by behavioural change, giving the striking example of the education programme run in the United States to change farming practices following the Dust Bowl. Bradsen describes this programme as the longest-running, most intensive effort ever undertaken to produce an outcome by attitudinal change. The programme cost some $30 million and yet has been judged a failure.

Similar criticisms could be, and indeed regularly are, made of the failure of all the 'techniques of thought and social organization' which Bradsen identifies. As Webber points out in Chapter 11, the model that 'science produces information, and that if and when the information is provided to land-users their decisions are rendered more fitting, is naive and does not stand up to careful examination'. It has also long been evident that the capitalist economic system, if left to itself, will not bring about the environmental change needed for a sustainable future. If this were not so, we would not have the problems we now have. Indeed the economic system is regularly identified as source of, rather than solution for, the problem. The law and the institutions established by law also have their manifest deficiencies. Disappointment with the efficacy of the US National Environmental Policy Act (1969) and the Environmental Protection Agency which it brought into being (which have been variously described as the most powerful piece of environmental legislation ever enacted and agency ever established) have led to despairing searches for alternative ways of bringing about more significant environmental change.[7]

On the other hand it is possible to point to significant, if limited, achievements which have been brought about by each of these means. Attitudinal change in Western democratic societies has been extremely influential in bringing about political responses to environmental problems. Science has been the main source of the identification of environmental problems and thus has been very influential in generating awareness and influencing attitudes. Science has also provided the technical answers needed to arrest environmental pollution or degradation. Economic incentives have at times proved very powerful tools for bringing about behavioural change. The law, as demonstrated by the examples cited by Bradsen, can also prove very effective.

In each of the four specific programmes referred to in this book, conservation farming, the salinity programme, Landcare and native vegetation protection, different combinations of these general groups of means can be identified. Science has played an important part in each. The law has been a powerful instrument for achieving native vegetation protection in South Australia. Education and attitude change through consensual processes have been core elements of the salinity programme and Landcare. Economic incentives are an important component of the salinity programme and, to a lesser degree, Landcare.

We might conclude that any successful environmental initiative will require a combination of good science and the intelligent use of at least one of the other groups of means. The South Australian native vegetation programme has worked through an effective use of science and the law. The salinity programme was successful because of its imaginative use of science, economic initiatives, consensual processes and administrative structure. Conservation farming has not been adopted widely because of the absence of adequate economic incentives or of deterrents to other less sustainable practices. Landcare has had only limited success because of the absence of significant economic incentives, the inadequate use of scientific knowledge (e.g. related to tree planting) and only partially effective consensual processes. Thus the more complex the problem, the more likely it is that a wide range of means needs to be called upon to achieve the desired environmental change.

The Perception of Appropriate Solutions

In the introduction to this book we argued that a characteristic of environmental problems is their complexity, and that this complexity takes many forms. In the preceding section it was shown that finding the right mix of solutions to environmental problems is also very complex. An approach based on a particular technique—the use of law, economic incentive, education or scientific input—or combination of techniques may work in one set of circumstances, but not in another. Often all of these techniques need to be used in concert to bring about behavioural change. Yet one of the striking features of proposals to achieve environmentally related change is their consistent reliance on the individual techniques with which the proponents are most familiar. This singularity of emphasis is equally apparent in this book. Scientists propose scientific solutions—better science, better transfer of scientific knowledge; agriculturalists and farmers suggest industry-related solutions—changes to new conservation farming techniques; lawyers suggest legal solutions, and so on. While each may elaborate very interesting ways in which the profession or activity with which he or she is familiar might more effectively be used, implicit in the silence about the other techniques is the suggestion that they are unimportant (or that the writer is unaware of their importance). Sometimes criticism of the other techniques, those other than the one that the writer or actor supports, is explicit.

Redefinition of Knowledge

Judith Innes has argued that the systematic failure to use information—here read scientific knowledge about the nature of land degradation—to solve problems can in part be attributed to the framework of ideas we hold about knowledge, the production of knowledge, and the role of the professional.[8] Most of our traditional assumptions about knowledge and the production of knowledge, she suggests, can be attributed to the positivist or scientific model. This model holds that

> knowledge consists in measurable facts that have an independent reality ... these facts may be discovered by a disinterested observer applying explicit replicable methodology ... knowledge consists in general laws and principles relating variables to one another ... such laws can be identified through logical deduction from assumptions and other laws and through testing hypotheses empirically under controlled external conditions.[9]

According to this model, genuine knowledge is about facts or abstract relationships, not about 'elusive' subjective attributes such as feelings and experiences. The generation of knowledge and decision-making are separate processes. The scientist or analyst collects the information. It is passed on to the decision-maker who acts on it.

Innes argues, as does Webber in this book, that what people actually do bears little resemblance to this model. An alternative model is therefore needed which is grounded in the everyday world, respects ordinary language and beliefs, and attempts to understand all the influences at play in a holistic way. Innes suggests that the model which best fits this description is a phenomenological or interpretive one. This model accepts that all research has its own biases and preconceptions: that its findings, if they are to be accepted as valuable and 'true', must reflect values and expectations in the community. As a consequence, much of a researcher's attention must be concerned with the choice of measures that will reflect the nature of the reality being studied. Expert knowledge cannot be the only privileged knowledge. Local, practical and intuitive forms of knowledge are as important as professional knowledge. Knowledge generation is an iterative process which alternates between objective intellectual activity and subjective experiential understanding.

Does the nature of knowledge generation in the form most useful for the solution of land degradation and for developing sustainable agricultural practices fit this model? The political processes described by Evan Walker in the development of the Victorian government's salinity programme, the problems identified by Leigh Walters and Albert Rovira which make it difficult for farmers to adopt new sustainable practices such as conservation farming, and the ways in which knowledge has been developed and exchanged in farmers' group sessions associated with the salinity and Landcare programmes all suggest a much better fit with this model than with the positivist scientific alternatives.

Transfer of Knowledge

In Chapter 13 Walters and Rovira pose two alternative models of knowledge transfer (see Figure 13.1, page 163). It can readily be seen that the linear model, Model A, reflects the positivist scientific model of knowledge production and transfer, and the interactive model, Model B, reflects the phenomenological or interpretive model of knowledge production and transfer. As Walters and Rovira observe, 'current thinking favours Model B in which the farmer, researcher and extension officer (or consultant) work through the problem together so that the farmer has a sense of ownership in the solution and is more likely to adopt the technology'. (Note the similarity between this pair of models and Faludi's 'object-centred' and 'decision-centred' models of planning, as quoted by Kwi-Gon Kim in Chapter 7.)

The Nature of Scientific Research and the Social and Political Nature of Environmental Problems

In Chapter 8 Ron Johnston and Tricia Berman illustrate why environmental research is not easily defined, and why, as a consequence, there are considerable disagreements about definitions. Their summary of the current status of and problems with environmental research is that

> it is a relatively new area of research, and is regarded by those working in traditional disciplines as being marginal. It is characterized by a focus on complex interdisciplinary problems. It is constantly being redefined as knowledge and techniques advance and as the significance of issues to be addressed changes.

To Brian Finlayson and Tom McMahon, the problems of Australian environmental science stem from a conceptual model of the natural environment and human relationships brought from Europe and North America that is inappropriate to Australia. They identify an excessive concentration of environmental research in government institutions which do not subject this research to peer assessment and review, either before or after the research is undertaken. Both groups of authors agree that the mechanisms for identifying and supporting environmental research are inadequate. To give point to the importance of directing environmental research more effectively, Johnston and Berman remind us that the population of Australia constitutes no more than 0.3 per cent of the world's population, while we have stewardship of 5 per cent of the world's land mass. Thus the money we can spend on understanding our natural environment is 'spread very thinly and must be spent to best effect'. The recommendations or possible means of improving environmental research offered by the four authors include:

- a research council dedicated to environmental research (as in the UK);
- an additional review panel in the Australian Research Council dealing only with interdisciplinary environmental research;
- the direction of more money to the universities (not competitively bid for) to encourage curiosity-driven research.

These views constitute an excellent summary of and insightful commentary on the problems of environmental research (as defined). The various recommendations proposed (only some have been referred to here), if adopted, would likely improve the focus of scientific environmental research significantly. The land-degradation case study supports the view that programmes are too often developed on the basis of an inadequate, outdated or unsubstantiated view of the nature of the scientific problems.

There are, however, some real difficulties with the underlying assumptions in the two chapters. The authors do not help us to understand how we might, given an improved scientific knowledge base, develop an interactive rather than linear pattern of knowledge transfer (as, for example, recommended by Walters and Rovira) which recognizes the importance of other forms of knowledge, including local, practical, political and feminist knowledge and understanding.

Even more significant is the general definition given to environmental research. All four authors seem to assume that environmental research means scientific research related to environmental problems. The Australian Science and Technology Council (ASTEC) definition of environmental research as quoted, 'work undertaken to acquire and organise knowledge of natural systems that sustain life in Australia ... knowledge [of] characteristics of natural systems, any impacts of human action on them and measures to integrate adverse impacts',[10] is also disturbingly narrow. If, as Joni Seager states in Chapter 10, 'environmental problems are quintessentially social and cultural problems—precisely the realm in which scientific understanding is at its weakest', how can we then define environmental research as solely or largely concerned with the physical sciences? There are problems for researchers in the physical sciences in getting support for multidisciplinary research across different realms of the physical sciences. How much greater, then, are the problems of researchers working in the social and political domains of environmental problems who have not even a category of research defined by the Australian Research Council to use in their applications?

In Chapter 11 Webber takes this argument further. He asserts that not only is environmental degradation fundamentally political; science is too. Science and land degradation are, moreover, two sides of the same coin. To develop this argument, Webber puts forward three propositions: 'internal and external limits constrain science to produce information that is inappropriate for controlling degradation; the significance of environmental science to public policy is grossly overrated by scientists; and some environmental interventions are tending to make environmental science irrelevant'. To what degree do the studies and other evidence offered in this book support these views?

In each of the four land-restoration programmes examined in this book, science and the market economy have demonstrably failed to find an answer to the environmental problem at issue, whether that problem concerned the introduction of a new conservation practice, the protection of native vegetation, or solutions to catchment-wide problems. In each instance, political intervention was required to bring about a more satisfactory outcome. The evidence thus strongly supports the view that land degradation is much more than a scientific problem; rather, it is inherently a political problem.

To support his statement that science produces knowledge that is inappropriate for the solution of problems caused by land degradation, Webber argues that, because knowledge about physical processes can be more easily assembled than knowledge about social processes, and knowledge about production more easily than knowledge about conservation, research inevitably has tended to concentrate on physical problems and production rather than on social processes and conservation (where it is most needed). These trends are reinforced by the lack of 'respectability' of locally specific research, which can be messy, and difficult to publish. The case study material implicitly supports this view. In Chapter 16 Evan Walker, for example, vividly describes the emotional and political nature of the salinity problem: 'The political realities are simple. Tough political decisions are almost impossible to take when seats are at stake.' The research needed in this instance was on the social and political nature of the problem.

Is the role of science overrated? The studies of the four programmes show that science was involved in every instance: to identify the potential of conservation farming as a counter-intuitive technique; to assess the value of stands of native vegetation as part of the procedures established under South Australian legislation for native vegetation retention; to identify the nature of salinity and technical solutions for its control for the salinity programme; and to identify the nature of overall catchment problems for the Landcare programme. To that extent, it could be argued that science has played a critical part. However, as has been demonstrated in this chapter, science and the market place alone have failed to solve any of the four groups of problems. So there is evidence to support Webber's view that the contribution of science to the solution of environmental problems is regularly overrated. This is not to say that science is unimportant. Science has a key role, but, as Joni Seager observes, it is only a tool, albeit a powerful and useful one.

Webber's final point is that

> once the human contribution has become more than trifling, the evolution of the new landscape is no longer natural, and catastrophes lose their status as simple acts of God. If there are floods or droughts, hurricanes or calms, hot weather or cold: all might be our effects. Science, having claimed to be the basis for our capacity to control nature and unleash technology for human benefit, will have to wear the blame.

It is a sobering argument.

Can science and technology control the forces they have unleashed? The increasing doubt that this is likely is reflected in Kwi-Gon Kim's chapter. In Korea, Kim argues, there is a growing view that the environmental problems caused by development based on Western science and technology will not be controlled by that science and technology alone. Hence the appeal to ancient Korean and Chinese principles and techniques, based on an intuitive spiritual relationship to the land, which might be used alongside Western science and technology.

Changes in Core Beliefs

In the literature on environmental attitudes and behaviour it has been argued that environmental problems can be attributed to traditional Western beliefs about progress, materialism, production and development and a divorce from nature. This traditional belief system has been called the Dominant Social Paradigm.[11] In contrast with these beliefs, a new set of environmentally conscious beliefs have emerged which support ecology, a limit to growth and development, and a life in harmony with nature. One formulation of this new belief system has been called the New Environmental Paradigm.[12] By this argument a sustainable future requires a progressive global change from the Dominant Social Paradigm to the New Environmental Paradigm.

It has also been argued that the environmental crisis can be seen as a crisis of maladaptive behaviour.[13] According to this view, current patterns of behaviour are inimical to the survival of the human species. Thus, what we should be concerned with is the process by which we achieve change towards environmentally adaptive behaviour. Implied is that the behaviour change required must be compatible with a changed environmental belief system such as the New Environmental Paradigm.

To examine the nature of environmental behavioural change, various conceptual models have been developed. The aim of the models is to identify as accurately as possible the major influences on behavioural change. Gray has, for example, discussed and compared models developed by Rokeach and by Fischbein and Ajzen.[14] In each model emphasis is placed on a hierarchy of beliefs. Of fundamental importance are primary beliefs, 'beliefs that form a part of a person's salient belief hierarchy'.[15] A formal principle of change in the Fishbein–Ajzen model, for example, is that the success of an attempt to bring about a change in attitude will depend on its effects on the primary beliefs underlying that attitude. Primary beliefs are deeply held. They will be altered only through a chain of effects starting with information that is likely to have substantive bearing on the beliefs and sometimes involving a series of intermediate steps.[16]

In Chapter 17 Sharman Stone discusses the long-established credo subscribed to by farmers in most developed countries which supports farmers' claims to a unique and special status (see page 205). While this credo may not directly correspond to the Dominant Social Paradigm as defined by Pirages and Ehrlich, it can readily be identified as part of a core belief system. Sharman Stone argues that these beliefs now constitute a serious barrier to the changes needed to develop new sustainable agricultural practices. As she points out, it has become increasingly apparent that 'individual actions can do little to arrest rising regional water tables, native forest dieback or the degradation of river systems'. If these problems can be solved only by co-operative action, then a different set of core beliefs is needed to sustain long-term change. The consensual processes which Stone describes when discussing the salinity and the Landcare programmes can thus be seen not only as processes which might lead to co-operative effort to solve immediate catchment-wide problems, but also as processes through which changes in farmers' core beliefs might take place.

Dialogue

As has been noted above, psychologists argue that environmental behavioural change, or indeed any other behavioural change, involves complex issues of underlying belief. Change will not occur through simple appeals to reason. The most effective way of achieving change to maladaptive core beliefs is to find opportunities for the confrontation of proximal beliefs with new information which will create dissonances between the beliefs and the information, thus causing those concerned to reflect carefully on the issues and to respond to peer group pressures.[17] These dissonances are ideally worked through in consensual processes.

Another way of examining this attitude–behaviour relationship is to look at it in the context of knowledge, policy, and action.[18] Participants in a joint task bring to it a variety of skills, knowledge, interests and experience. When effective communication takes place between the participants, there is an exploration, through debate and negotiation, of definitions, values and common understandings which can lead to new courses of action. Debates and negotiation and new actions in their turn can lead to changes of perception and understanding. Within this process several forms of knowledge may be required which reflect different kinds of interests, technical, practical and critical. Each of these forms of knowledge has its place: the technical or scientific to solve problems; the practical to understand what is happening, to communicate and to bring about action; the critical to uncover the hidden implications of action and to challenge current processes and assumptions. These forms of knowledge are best brought into play through combined forms of communication and action. The development of thinking about these processes owes much to the work of Jurgen Habermas and his theory of communicative action.[19]

There are linkages between the Fischbein–Ajzen theory of belief systems and the larger theoretical framework with which Habermas is concerned. Well-designed consensual processes and forms of communicative action can lead to new solutions for immediate problems. At the same time they can bring about changes in understandings, proximal beliefs and eventually core beliefs. Through these changes in core beliefs, sustained long-term change can be achieved.

Sharman Stone's description of the farmers' involvement in the salinity programme is a model of this process. A very high percentage of farmers participated, and by working through the issues together, confronting their various assumptions and preconceptions, and using the knowledge of outside experts, were able to agree upon a range of co-ordinated actions. The salinity programme is thus a fine illustration of a consensual process or form of communicative action working well. What it and Landcare show, however, is that the critical issue is the way the process is planned and organized. The

consensual structure in both the salinity and Landcare programmes was identical. In one, very high involvement was achieved;
in the other, involvement has tended to be relatively low. The
key to success can thus be seen to be the design of the process.
There must be incentives to get participants involved (in the
salinity programme funds were available only when catchment-
wide co-operative agreement could be reached). There must
also be deterrents against failure to participate (in the salinity
programme the fear of other farmers' actions). Finally, there
must be expert help and sufficient time. Time is a particularly
important requirement if, through these actions and the continuing discussions, explorations and confrontations of preconceptions, a progressive change in core beliefs is to take place.
Without that change it is likely that many farmers will revert to
old non-sustainable habits.

Conclusion

In this book there is evidence to support each of the propositions with which this chapter began. Material in the book,
furthermore, supports the underlying theme of the book, the
pervasive influence of values. Values shape the way we define
terms, assess problems and choose courses of action. To achieve
a sustainable future, we need to find and adopt a new shared
environmental ethic which recognizes the critical influence of
values on environmental decision-making. The hope that is
demonstrated in this book is that so many different people,
representing so many diverse strands of interest and background—philosophers, theologians, scientists, geographers,
educationalists, politicians, farmers, journalists, lawyers, soci-
ologists, planners—are all struggling to define in their own
terms what that ethic should be.

One additional theme, strongly expressed in this book but
not directly reflected in the case study, concerns the renewed
attempt to define a human relationship with the spirit of the
land. Eric Willmot asks for 'a new mind, a kind of third
humanity'. Peter Hollingworth argues for a creative and co-
operative relationship with nature. Kwi-Gon Kim discusses the

revival of P'ungsui in Korea as both a popular and intellectual
movement seeking a reconnection with an ancient tradition of
spiritual relationship of human beings and the land.

Freya Mathews' story of the quandong tree—'I realized with
a shock that this was the first time that I had ever eaten of the
wild, native food of this land'—is a telling example of the alienation of white Australians from the land. She reminds us that
our refusal to partake of the land is 'a refusal to connect with
its essential nature, its spirit; if you like, its *genius loci*'. Aboriginal
poetry and painting illustrate Mathews' point perfectly. It would
be a surprise to find a white Australian poem about waterlily
roots as a source of food without mention of the waterlily flower.
'*A Bushtucker Story*' or '*Seedcake Dreaming*' would be unlikely titles
for a white Australian painting.[20]

How might we find a new relationship with the land, a
re-enchantment of our world, as Morris Berman has described
it?[21] The sharing of common understandings, communicative
action between the first and later dwellers, between those who
live on the land and those who live in the city, as Peter Small
pleads, might be one way. Another might be through the eyes,
ears, hands and voices of Australian artists. Mathews begins her
paper with a poem by Judith Wright. A fitting way to end this
book might again be with Judith Wright:

> the poet is, *par excellence*, the speaker for natural or biological
> man. This is the part of us which is least under conscious control,
> which feels and has emotions, rather than thinks and analyses:
> . . . Poetry, music, painting, are rooted far back in human history
> and in the natural rhythms of life. Even today, when most of us
> have forgotten those rhythms, art still employs them—which is
> why music, pictures and rhythmic verse and dance often move us
> in ways we hardly recognise. They are based on our biological
> cycles, which we hardly notice now, caught as we are in the stop–
> go, walk–don't walk, nine-to-five, artificial timetables of our lives.
> But poets listen to them still, which is why biological man can get
> through to poets more easily than to politicians and
> industrialists.[22]

Notes

1 Introduction

[1] For an account of this, see David Yencken, 'The links between issue identification and action', pp. 253–71.
[2] See, for example, Doug Miller, 'What the polls tell us', pp. 68–9.
[3] Ishimure Michiko, *Paradise in the Sea of Sorrow: Our Minamata Disease.*
[4] Helen Briassoulis, 'Theoretical orientations in environmental planning', pp. 381–92.
[5] L. E. Woods, *Land Degradation in Australia.*
[6] Brian Roberts, *Land Care Manual.*

Part I Values: The Meaning of the Land

[1] M. D. Williams, *Out of the Mist, Book II: The Story of Man's Mastery of his Physical Environment, and his Intellectual and Spiritual Development*, p. iv.
[2] Williams, *Out of the Mist*, p. x.
[3] Peter Knudtsen and David Suzuki, *Wisdom of the Elders.*
[4] See, for example, Fritjof Capra, *The Tao of Physics.*
[5] O. B. Hardison, *Disappearing through the Skylight: Culture and Technology in the Twentieth Century*, p. 5.

2 The Futility of Utility

[1] Barron Field, *Geographical Memoirs of New South Wales*, p. 254.
[2] Lionel Wigmore, *Struggle for the Snowy*, p. 194.
[3] Denys Blakeway and Sue Lloyd-Roberts, *Fields of Thunder*, p. 124.
[4] Debates over land rights have erupted most recently since the Australian High Court decision on the Mabo case in June 1992, which recognized that Aboriginal land title had not necessarily been extinguished by European settlement.
[5] Relevant arguments about humanism, anthropocentrism, and green ethics can be found in David Ehrenfeld, *The Arrogance of Humanism*; Christopher Manes, *Green Rage: Radical Environmentalism and the Unmaking of Civilization*;

Roderick Frazier Nash, *The Rights of Nature: A History of Environmental Ethics*; Christopher D. Stone, *Earth and Other Ethics: The Case for Moral Pluralism*; and Colin Tudge, *Last Animals at the Zoo: How Mass Extinctions Can Be Stopped.*

3   A New Mind and a New Earth

[1] G. N. Seagram and R. J. Lendron, *Furnishing the Mind: Comparative Study of Cognitive Development in Central Australian Aborigines.*
[2] Other works of relevance include I. Illich, *De-Schooling Society*, and E. Willmot, *The Culture of Literacy; Australia, the Last Experiment*; and *Education's Hidden Outcome.*

4   Values Implied by the Doctrine of Creation

[1] Daniel C. MacGuire, 'Values'.
[2] *An Australian Prayer Book*, p. 162.
[3] See, for example, W. Brueggeman, *The Land: Overtures to Biblical Theology*; S. McDonagh, *To Care for the Earth: A Call to a New Theology*; and G. R. Lilburne, *A Sense of Place: A Christian Theology of the Land.*
[4] J. Moltmann, *God in Creation: An Ecological Doctrine of Creation.*
[5] Moltmann, *God in Creation*, p. 30.
[6] The New Revised Standard Version Bible.
[7] Moltmann, *God in Creation*, preface, p. xi.

5   *Terra Incognita*: Carnal Legacies

[1] Judith Wright, 'Eroded Hills', in *Collected Poems*, p. 83. My thanks to David Tacey for drawing my attention to this and other poems.
[2] Although mind and body were not sharply divided in the art and literature of classical Greece, and the culture generally was imbued with an aesthetic appreciation of the human body and the natural world, the seeds of 'dualistic' thinking with respect to mind and body can be found in Greek philosophy, particularly in the work of Plato and Aristotle. Again, neither of these thinkers presents mind and body in sharply dichotomous terms, but Plato does divide human nature into three components: reason, 'spirit' (analogous to 'fighting spirit' or the spiritedness attributed to animals), and appetite; and he sees reason as the 'higher' component, properly ruling 'spirit' and the 'lower' bodily appetites. According to Aristotle, reason is the essence of human nature: it is reason which makes us human. For a feminist analysis of the development of a dualistic notion of reason in the Western tradition, see Genevieve Lloyd, *The Man of Reason.*
[3] There is a large and diverse literature tracing the influence of Christianity on Western attitudes to Nature. The *locus classicus* for the case against Christianity is Lynn White Jnr, 'The historical roots of our ecological crisis', pp. 1203–7. A number of Christian apologists argue that the Judeo-Christian scriptures contain several possibilities in this connection—the 'dominion over Nature' tradition can be offset by the 'stewardship' tradition, for instance (see Peter Hollingworth in Chapter 4 of this book). Mathew Fox is perhaps the best-known theologian at present exploring ecological possibilities within the creation spirituality tradition of Christianity (see M. Fox, *Original Blessing*). See also Geoffrey Lilburne, *A Sense of Place: A Christian Theology of the Land.*

⁴ For a summary of a wide literature on this topic, see Freya Mathews, *The Ecological Self.*
⁵ Mathews, *The Ecological Self.*
⁶ Some of these developments are also discussed in Mathews, *The Ecological Self.*
⁷ Primary exponents of Deep Ecology include Arne Naess, 'The shallow and the deep, long-range ecology movement', and *Ecology, Community and Lifestyle*; Bill Devall and George Sessions, *Deep Ecology: Living as if Nature Mattered*; and Warwick Fox, *Towards a Transpersonal Ecology.* Introductory readings in ecofeminism include Judith Plant (ed.), *Healing the Wounds: The Promise of Ecofeminism*; J. Diamond and G. Orenstein (eds), *Reweaving the World.* For an overview of the various ecophilosophies, see Carolyn Merchant, *Radical Ecology.*
⁸ Veronica Brady makes this observation in 'Called by the land to enter the land'.
⁹ Alfred W. Crosby makes the point in *Ecological Imperialism: the Biological Expansion of Europe, 900–1900* that the European invasion of this—and other—lands was a biological as well as a human invasion: we brought with us all our biological baggage, including our diseases.
¹⁰ A statement of the identification of indigenous peoples with the native foods of their lands is made by Gwaganad of Haada Gwaii (the Queen Charlotte Islands) in her testimony to the Supreme Court of British Columbia in 1985 over a land rights issue. She is speaking of the way her body feels at the time of the herring spawn: 'I get a longing to be on the sea . . . My body is kind of on edge in anticipation . . . Finally the day comes when [the herring] spawns . . . And you don't quite feel complete until you are right out in the ocean with your hands in the water harvesting the kelp . . . your body almost rejoices in that first feed. In order to make me complete I need the right food from the land . . . I have to harvest it myself'. See 'Speaking for the earth: the Haida way'.
¹¹ Of course the boundary was not always, or even typically, sharp: peoples who practised horticulture also foraged in nearby forests and elsewhere for food, fodder, fuel and other goods. However, the sense of a specifically human space, on the one hand, and a space beyond, on the other, undoubtedly accompanied the advent of cultivation, even when the space beyond was not entirely inaccessible to humans.

6   Property Rights and the Environment
¹ W. N. Hohfeld, *Fundamental Legal Conceptions as Applied in Legal Reasoning*, p. 28.
² The Australian High Court's recognition of Aboriginal land title in the Mabo case may be seen as an acknowledgement of alternative conceptions of ownership and property rights.
³ Andrew Reeve, *Property*, p. 13.
⁴ Stanley Benn, 'Property', pp. 491–4; Hohfeld, *Fundamental Legal Conceptions*; and A. M. Honoré, 'Ownership', pp. 107–47, analyse the concept of property as sets of rights and liabilities.
⁵ Honoré, 'Ownership', p. 108.
⁶ Honoré includes both rights and responsibilities in his list of the standard features of ownership, so he refers to these features as the 'incidents of property' rather than as 'property rights'.

[7] Honoré, 'Ownership', p. 113. The right to transmit property is the right to give or lend property rights over an object. Absence of term indicates that an owner may enjoy her property rights for an unlimited duration, unless she chooses to sell or give away her rights. Residuarity refers to the revival of an owner's property rights when, for example, a lease is terminated. Honoré does not claim that this is the only way in which the standard incidents of ownership could be enumerated (p. 113). Further division could occur among, for example, the various rights which constitute the right to capital, or the rights of use and management could be collapsed. His point is to show the richness of property rights, not to exhaust the possibilities.

[8] In the sense described by Waldron as a system based on the organizing idea of private ownership; Jeremy Waldron, 'What is private property?', pp. 326–33.

[9] Robert Nozick and Murray Rothbard are two well-known proponents of the libertarian view—see Nozick, *Anarchy, State, and Utopia*; and Rothbard, *For a New Liberty*. A less extreme liberty-based theory of property rights can be found in the work of John Locke, *Two Treaties of Government*.

[10] See Nozick, *Anarchy, State, and Utopia*, pp. 79–81.

[11] A 'shallow' ecological ethic is based on traditional human-centred ethical concepts which have been refined in the light of awareness of our dependence on the environment. 'Deep' ecology, on the other hand, is grounded in ecocentric, communal ethical concepts. See C. A. Hooker, 'Responsibility, ethics and nature', pp. 147–64.

[12] Robert Elliot, 'Future generations, Locke's proviso and libertarian justice', pp. 217–27.

[13] While most of the discussion has been focused on limits to private property rights, the analysis applies equally strongly to communal or state property. If the justification for making the state responsible for large tracts of bushland is that the state is best placed to efficiently manage the land for the enjoyment and benefit of all members of the community, then it may be obliged to preserve the natural features of the land. If the local council is given exclusive control of hydro-electric power plants on the grounds of utility, then the council can be held accountable for ensuring that the running of those facilities does not harm others, and that legitimate environmental values are respected in the exercise of rights over the plant.

[14] Cf. C. B. Macpherson, *The Political Theory of Possessive Individualism: Hobbes to Locke*.

7 Learning from Other Cultures

[1] A. Faludi, 'Three paradigms of planning'.

[2] See L. Ortoland, *Environmental Planning and Decision Making*, p. 5; J. Passmore, *Man's Responsibility for Nature: Ecological Problems and Western Traditions*; L. H. Tribe, 'Ways not to think about plastic trees', pp. 61–92.

[3] M. Bring and J. Wayembergh, *Japanese Gardens: Design and Meaning*, p. 2.

[4] Kwi-Gon Kim, *Report of Working Group Two on Urbanization and Environmental Change*.

[5] Sang-Don Lee, Kwi-Gon Kim and Sang-Gon Lee, *National Report of the Republic of Korea to United Nations Conference on Environment and Development*.

[6] J. O. Simonds, *Landscape Architecture: The Shaping of Man's Natural Environment*, pp. 76–7.

⁷ J. K. Koh, 'Problems in the New Town Development Plan in Korea and a call for human ecological approach', pp. 62–8. See also Ja-Hae Lim, 'Tradition in Korean society: Continuity and change', p. 24.
⁸ C. J. Choi, 'The P'ungsui theory (geomancy) and a Korean view of land', p. 34.
⁹ See F. Capra, *The Tao of Physics*, and F. Capra and D. Steindl-Rast, *Belonging to the Universe.*
¹⁰ D. E. Mann, 'Environmental learning in a decentralized political world', pp. 330–1.

Part II   Knowledge: Asking the Right Questions
¹ See, for example, K. E. F. Watt, *Understanding the Environment*, pp. 340–57.

8   Environmental Research Policy
¹ Cabinet Office Advisory Council on Science and Technology (ACOST), *Environmental Research Programmes*, p. 1.
² ASTEC, *Environmental Research in Australia: A Review*, p. 2.
³ M. Schaefer, 'The federal research puzzle: Making the pieces fit', p. 18.
⁴ Australia, *Ecologically Sustainable Development: A Commonwealth Discussion Paper.*
⁵ R. Bradbury, in a paper presented to the ASTEC Environmental Research Seminar held at the Australian National University, Canberra, 12 June 1991. See also note 12 below.
⁶ ASTEC, *Environmental Research in Australia: Case Studies*, p. 108.
⁷ ACOST, *Environmental Research Programmes.*
⁸ Resource Assessment Commission, *Kakadu Conservation Zone Inquiry* and *Forest and Timber Inquiry.*
⁹ Biological Diversity Advisory Committee, *A National Strategy for the Conservation of Australia's Biological Diversity: Draft for Public Comment.*
¹⁰ Economic Planning Advisory Council (EPAC), *Managing Australia's Natural Resources*, p. 39.
¹¹ ASTEC, *Environmental Research in Australia: Case Studies*, pp. 89–115.
¹² These issues were raised at the ASTEC Environmental Research Seminar held at the Australian National University, Canberra, 12 June 1991, to discuss the recommendations of the ASTEC review of environmental research in Australia (ASTEC, *Environmental Research in Australia: The Issues*; and ASTEC, *Environmental Research in Australia: Case Studies*).
¹³ Organisation for Economic Co-operation and Development (OECD), Directorate for Science, Technology and Industry, *Environmental Change and Science and Technology Institutions, The Experience of Selected OECD Countries*, p. 50.
¹⁴ CSIRO is the Commonwealth Scientific and Industrial Research Organisation; AIMS is the Australian Institute of Marine Science; ANSTO is the Australian Nuclear Science and Technology Organisation; DSTO is the Defence Science and Technology Organisation; NH&MRC is the National Health and Medical Research Council; BMR is the Bureau of Mineral Resources; BRR is the Bureau of Rural Resources.
¹⁵ The Hon. Alan Griffiths, Opening address to 1990 AURISA Conference, Canberra.
¹⁶ ASTEC, *Environmental Research in Australia: The Issues*, p. 11.
¹⁷ ESD Working Group Chairs, *Intersectoral Issues Report*, p. 232.

[18] Australian Academy of Science, *Global Change: A Research Strategy for Australia, 1992–96.*

9   Funding and Conduct of Environmental Research

[1] T. A. McMahon *et al.*, *Global Runoff: Continental Comparisons of Annual Flows and Peak Discharges.*
[2] ASTEC, *Environmental Research in Australia: A Review.*
[3] N. Porter (ed.), *Webster's International Dictionary.*
[4] ASTEC, *Environmental Research in Australia: The Issues*, p. 5.
[5] ASTEC, *The Issues*, p. 59.
[6] ASTEC, *A Review*, p. 21.
[7] ASTEC, *A Review*, p. 21.
[8] ASTEC, *A Review.*
[9] ASTEC, *A Review*, p. 21.
[10] T. A. McMahon and B. L. Finlayson, 'Australian surface and groundwater hydrology—regional characteristics and implications', pp. 21–40.
[11] J. W. Porter and T. W. McMahon, *The Monash Model: User Manual for Daily Program HYDROLOG.*
[12] F. H. S. Chiew and T. A. McMahon, 'Improved modelling of the groundwater process in HYDROLOG', pp. 492–7.
[13] F. H. S. Chiew *et al.*, 'Estimating groundwater recharge using an integrated surface and groundwater modelling approach', pp. 151–86.
[14] Land Conservation Council, *Rivers and Streams: Special Investigation.*
[15] B. L. Finlayson and J. F. Bird, *Initial Investigation into the Extent and Nature of the Current Sedimentation Problem on the Lower Snowy River.*
[16] F. P. Larkins, 'Science and technology policies in Australia: Planning for the future'.
[17] Centre for Policy Research, *University Research Infrastructure: An Indicative Study.*
[18] Murray–Darling Basin Ministerial Council, *Murray–Darling Basin Natural Resources Management Strategy.*
[19] F. Jackson, 'How not to fund research', pp. 16–18.

10   Environmental Research: A Feminist Critique

[1] See, for example, Sandra Harding, *The Science Question in Feminism*; Evelyn Fox Keller, *Reflections on Gender and Science*; and Ruth Bleier (ed.), *Feminist Approaches to Science.*
[2] Evelyn Fox Keller, 'Contending with a masculine bias in the ideals and values of science', p. 96.
[3] Carolyn Merchant, *The Death of Nature.*
[4] See, for example, Donna Haraway, 'Animal sociology and a natural economy of the body politic', pp. 21–60.
[5] See Chapter 4 by Peter Hollingworth, and Chapter 11 by Michael Webber.

11   Politics, Science and the Control of Nature

[1] See K. Boardman and R. Eckersley, 'An Australian agenda', pp. 25–57; and W. D. Williams, 'Water for sustainable resource management within a semi-arid continent', pp. 11–17.
[2] J. Young, *Post Environmentalism*, pp. 88–9.

[3] I. Spiegel-Rosing, 'The study of science, technology and society: recent trends and future challenges', pp. 7–42.
[4] Department of Employment, Education and Training, *Australian Research Grants Scheme: Grants and Fellowships Awarded 1988*, p. 8.
[5] J. Irvine, 'Australian government funding of academic and related research—the international comparison', pp. 88–95.
[6] Organisation for Economic Co-operation and Development (OECD), *Reviews of National Science and Technology Policy: Australia*, p. 90.
[7] Australian Science and Technology Council (ASTEC), *Setting Directions for Australian Research*.
[8] See R. MacLeod, 'Changing perspectives in the social history of science', pp. 148–95, and B. Martin, *Scientific Knowledge in Controversy*, pp. 149–55.
[9] M. Mortimore, *Adapting to Drought*.
[10] M. J. Mulkay, 'Sociology of the scientific research community', pp. 93–148.
[11] D. de S. Price, 'Is technology historically independent of science?', pp. 553–67.
[12] E. Layton, 'Conditions of technological development'.
[13] See D. Bloor, 'Durkheim and Mauss revisited; classification and the sociology of knowledge', pp. 51–75; and P. Forman, 'Kausalitat, Anschaulichkeit and Individualitat', pp. 333–47.
[14] Mulkay, 'Sociology', pp. 100–4.
[15] R. Dumsday, 'Contributions from the social sciences', pp. 315–34.
[16] C. Chartres, 'Australia's land resources at risk', pp. 7–26.
[17] See R. Wasson, 'Detection and measurement of land degradation processes', pp. 49–69; and G. Robertson, 'Contributions from the physical and biological sciences', pp. 305–14.
[18] M. J. Kirkby, 'The problem', pp. 1–16.
[19] See D. I. Smith and B. L. Finlayson, 'Water in Australia: its role in environmental degradation', pp. 7–48; and B. G. Williams, 'Salinity and waterlogging in the Murray–Darling Basin', pp. 87–120.
[20] See Dumsday, 'Contributions from the social sciences', pp. 315–34; S. Neville, *The Australian Environment: Taking Stock and Looking Ahead*; and I. J. Reeve, R. A. Patterson and J. W. Lees, *Land Resources: Training Toward 2000*.
[21] See Wasson, 'Detection and measurement'. For physical controls see J. K. Mitchell and G. D. Bubenzer, 'Soil loss estimation', pp. 17–62; and N. P. Woodruff and F. H. Siddoway, 'A wind erosion equation', pp. 602–8.
[22] L. E. Woods, *Land Degradation in Australia*, Tables 5.1 and 7.3.
[23] Parliament of Victoria Salinity Committee, *Salt of the Earth: Final Report on the Causes, Effects and Control of Land and River Salinity in Victoria*, pp. 16–41.
[24] See Smith and Finlayson, 'Water in Australia', p. 27, and J. J. Jenkin, 'Salinity problems in Australia', pp. 141–52.
[25] G. Burch, D. Graetz and I. Noble, 'Biological and physical phenomena in land degradation', pp. 27–48.
[26] R. W. Galloway, 'Natural and anthropogenic erosion and their relative impact in the arid zone', pp. 45–6.
[27] R. W. Condon, 'Pastoralism', pp. 54–60.
[28] See J. A. Mabbutt, *Desertification in Australia*; and Neville, *The Australian Environment*.
[29] Galloway, 'Natural and anthropogenic erosion', pp. 45–6.
[30] R. J. Stanley, 'Soils and vegetation: an assessment of current status', pp. 8–18.
[31] M. Blyth and A. McCallum, 'Onsite costs of land degradation in agriculture and forestry', pp. 79–98.

[32] R. Rickson *et al.*, 'Social bases of farmers' responses to land degradation', pp. 187–200.

[33] Blyth and McCallum, 'Onsite costs', pp. 79–98.

[34] G. Upstill and T. Yapp, 'Offsite costs of land degradation', pp. 99–109.

[35] See J. Bradsen and R. Fowler, 'Land degradation: legal issues and institutional constraints', pp. 129–67; and Dumsday, 'Contributions from the social sciences', pp. 315–34. Susan Dodds discusses this in some detail in Chapter 6 of this book.

[36] See Blyth and McCallum, 'Onsite costs', pp. 79–98; and J. Emel and R. Peet, 'Resource management and natural hazards', pp. 49–76.

[37] H. Dorn, *The Geography of Science.*

[38] See for example p. Singer, 'Environmental values', pp. 3–24; and J. Young, *Post Environmentalism.* However, Peter Hollingworth argues in Chapter 4 of this book that this is a misreading of Christian theology.

[39] Major-General Thomas Sands, US Army Corps of Engineers, quoted in J. McPhee, *The Control of Nature*, p. 108.

[40] D. G. Smith, *Continent in Crisis*, p. 123.

[41] Young, *Post Environmentalism.*

[42] See for example M. Bookchin, *Towards an Ecological Society*, and *The Modern Crisis*; A. Naess, and G. Sessions, 'Basic principles of deep ecology', pp. 3–7; and the essays in M. E. Zimmerman, *Environmental Philosophy.*

[43] T. O'Riordan, 'The challenge for environmentalism', pp. 77–102.

[44] R. P. C. Morgan, 'Implications', pp. 253–301.

[45] P. Blaikie, *The Political Economy of Soil Erosion in Developing Countries.*

[46] Burch *et al.*, 'Biological and physical phenomena', pp. 46–8.

[47] Chartres, 'Australia's land resources', p. 24.

[48] E. K. Christie (ed.), *Desertification of Arid and Semiarid Grazing Lands*, pp. 1–5.

[49] B. T. Hart, 'Water quality management principles', pp. 1–5.

[50] Jenkin, 'Salinity problems in Australia', pp. 141–52.

[51] Morgan, 'Implications', pp. 253–301.

[52] Smith and Finlayson, 'Water in Australia', pp. 42–4.

[53] Williams, 'Salinity and waterlogging', pp. 87–120.

[54] Victorian Parliamentary Salinity Committee, *Salt of the Earth.*

[55] P. R. Hartley and M. G. Porter, 'A green thumb for the invisible hand', pp. 243–68.

[56] P. Ehrlich, *The Population Bomb.*

[57] Blyth and McCallum, 'Onsite costs', pp. 79–80.

[58] Department of Environment, Housing and Community Development, *Commonwealth and State Government Collaborative Soil Conservation Study 1975–77 Report 1: A Basis for Soil Conservation Policy in Australia.*

[59] Department of Primary Industries and Energy, *Third National Conservation Report.*

[60] Upstill and Yapp, 'Offsite costs', pp. 99–109.

[61] See for example Blyth and McCallum, 'Onsite costs'; Clarke et al. *Immigration, Population Growth and the Environment*; B. Davidson, 'Comments'; pp. 357–62; Dumsday, 'Contributions from the social sciences', pp. 315–17; Hartley and Porter, 'A green thumb', p. 244.

[62] Chartres, 'Australia's land resources'; Galloway, 'Natural and anthropogenic erosion'; Neville, 'The Australian environment'; Robertson, 'Contributions from the physical and biological sciences'; Smith, 'Continent in crisis'; Smith and Finlayson, 'Water in Australia'; Williams, 'Salinity and waterlogging'.

[63] B. Martin, *Scientific Knowledge in Controversy.*

[64] J. Alexandra, 'Political, social, economic and environmental issues in Australian water management', pp. 111–21.

[65] See Rural Water Commission of Victoria, *Inquiry into Water Allocations in Northern Victoria*; J. Eberhardt *et al.*, *Economic and Financial Issues*, *Water 2000*; and J. Keary, 'Victoria's new approach to water development', pp. 135–41.

[66] Eberhardt *et al.*, *Economic and Financial Issues*, pp. 119–20; and W. D. Watson *et al.*, *Agricultural Water Demand and Issues: Water 2000*, p. 15.

[67] Eberhardt *et al.*, *Economic and Financial Issues*, p. 167.

[68] I. R. McPhail and E. M. Young, 'Water for the environment in the Murray–Darling Basin', pp. 191–210.

[69] See, for example, S. A. Lakoff, 'Scientists, technologists and political power', pp. 355–91; and E. D. Ongley, 'Information requirements for water quality management: a reflective appraisal of present practices and future requirements', pp. 7–21.

[70] J. R. Ravetz, 'Criticisms of science', pp. 71–89.

[71] Lakoff, 'Scientists, technologists and political power', p. 381.

[72] Martin, *Scientific Knowledge in Controversy*.

[73] B. Jameson, *Movement at the Station*.

[74] B. McKibben, *The End of Nature*.

[75] R. G. Kazmann, civil engineer from LSU, quoted in McPhee, *The Control of Nature*, p. 143.

[76] McKibben, *The End of Nature*, pp. 88–124.

Part III   Action: Dealing with Land Degradation

[1] L. E. Woods, *Land Degradation in Australia*.

12   Sustainability of Agricultural Land

[1] See Chapter 7 of this book, by Kwi-Gon Kim.

[2] N. C. Uren, 'The management of soil organic matter for sustainable agriculture' pp. 45–8.

[3] A. R. Egan, 'Animal production—leading the recovery', pp. xvii–xxi.

[4] P. Buringh and H. D. Heemst, 'Potential world food production', pp. 19–72.

[5] Food and Agriculture Organisation, *Energy and Protein Requirements: Report of Joint Expert Committee*.

[6] This discussion of carrying capacity does not take into account the present inequitable distribution of the world's food, or the possibility that future social change might exacerbate this situation. The supply of water and of oil to power agricultural equipment may also prove to be limiting factors in food production. The point to note is that, despite these unknowns, external inputs such as fertilizers, oil and irrigation will be necessary if the expected future population of the world is to be fed.

[7] R. S. Loomis and D. J. Connor, *Crop Ecology: Productivity and Management in Agricultural Systems*, p. 538.

13   Turning Research into Action

[1] C. M. Donald and J. A. Prescott, 'Trace elements in Australian crop and pasture production, 1924–1974', p. 7.

[2] R. J. Hannam and D. J. Reuter, 'Review: Trace element nutrition of pastures', p. 175.

[3] Standing Committee on Agriculture, *Sustainable Agriculture*, p. 22.

[4] F. T. Hurley, B. C. Fitzgerald, J. T. Harvey and P. P. Oppenheim, *Cropping and Conservation: A Survey of Cultivation Practices in Victorian Grain Growing Areas*, p. 69.

[5] R. J. French and J. E. Schultz, 'Water use efficiency of wheat in a Mediterranean-type environment, II: Some limitations of efficiency', pp. 765–75.

[6] These criteria of sustainability apply to the farm itself. If the system were to be extended to include the means of providing external inputs such as fuels and fertilizers, the future supply of these would also have to be considered.

14   The View from the Farm

[1] A. B. Walker, 'Responses in the New Zealand meat and wool sector downturn'.

[2] Charles Massy, *The Australian Merino*.

[3] W. Crane, 'The role of trees in sustainable agriculture'.

[4] Brian Finlayson and Tom McMahon have more to say on this in Chapter 9 of this book.

[5] N. F. Barr and J. W. Cary, *Greening a Brown Land: The Australian Search for Sustainable Land Use*.

[6] A. R. Egan, 'Animal production—leading the recovery', pp. xvii–xxi.

15   Alternatives for Sustainable Land Use

[1] William Lines and Eric Willmot work through some of the implications of this assertion in Chapters 2 and 3 of this book, respectively.

[2] V. G. Carter and T. D. Dale, *Topsoil and Civilization*; D. E. Gelburd, 'Managing salinity: Lessons from the past', pp. 329–31; J. Reader, *Man on Earth*.

[3] Bryan G. Norton, *The Preservation of Species*.

[4] Stephen Garnett (ed.), *Threatened and Extinct Birds of Australia*.

[5] E. O. Wilson (ed.), *Biodiversity*.

[6] Fritjof Capra, *The Turning Point: Science, Society and the Rising Culture*.

[7] See, for example, B. R. Roberts, 'Land ethics—A necessary addition to Australian values'; and Aldo Leopold, *A Sand County Almanac*.

[8] J. R. Bradsen, *Soil Conservation Legislation in Australia*, pp. 83–99.

[9] S. Wittwer, 'New technology, agricultural productivity and conservation', pp. 201–15; W. D. Rasmussen, 'History of soil conservation institutions and incentives', pp. 3–18; L. K. Fisher, 'Discussion of trade-offs among soil conservation, energy use, experts and environmental quality', pp. 274–9.

[10] S. B. Lovejoy and T. L. Napier, 'Conserving soil: Sociological insights', p. 306.

[11] Herman E. Daly and John B. Cobb Jr, *For the Common Good*.

[12] Mark Sagoff, *The Economy of the Earth*.

[13] A. Chisholm, 'Abatement of land degradation: Regulations vs. economic incentives', pp. 223–47.

[14] J. R. Bradsen, 'Hard sticks or soft carrots'. See also Susan Dodds, in Chapter 6 of this book.

[15] The courts clearly support the legislation, but in a criminal law context it is simply too complex and cumbersome: *Woodbury County Soil Conservation District v. Ortner* (1979) NW 2d 276 USA.

[16] J. R. Bradsen, 'Perspectives on land degradation', pp. 16–40.

[17] These issues are summarized in Bradsen, 'Perspectives on land degradation'.

[18] This is explained in several articles in the *Ausralian Journal of Soil and Water Conservation*. It is put in its international context by John Allwright, 'The environment and sustainable growth—the key role of farmers', pp. 4–6.

¹⁹ Commonwealth of Australia, *Decade of Landcare Plan: Commonwealth Component*; ACT Parks and Conservation Service, *ACT Decade of Landcare Plan: Landcare a Project for All*; Adelaide Soil and Water Conservation Branch, *Decade of Landcare Plan for South Australia: Towards Sustainable Land Resource Management*; Department of Conservation and Environment, *The Victorian Decade of Landcare Plan*; Conservation Commission of Northern Territory, *Northern Territory Decade of Landcare Plan*; Department of Agriculture, and Soil and Land Conservation Council, *Decade of Landcare Plan, Western Australia*; Department of Primary Industry and Fisheries, *The Tasmanian Decade of Landcare Plan.*
²⁰ Australian Soil Conservation Council, *National Soil Conservation Strategy.*
²¹ Murray–Darling Basin Commission, *Managing Australia's Heartland.*
²² Biological Diversity Advisory Committee, *A National Strategy for the Conservation of Australia's Biological Diversity.*
²³ Allwright, 'The environment and sustainable growth'.
²⁴ Western Australian Parliament, Legislative Council, Debates, 1945, p. 562.
²⁵ *Soil Conservation and Land Care Act*, see Sec. 38; see also the definition of degradation in Sec. 3, and the duty of landholders in Sec. 8.
²⁶ *Native Vegetation Act* 1991, Schedule 1.
²⁷ The author chairs the Native Vegetation Council.

16   Government Policy and Environmental Change

¹ Victorian Parliamentary Salinity Committee, *Salt of the Earth.*
² Victorian Parliamentary Salinity Committee, *Salt of the Earth.*
³ Victoria, Government, *Salt Action–Joint Action.*
⁴ Sharman Stone gives an account of how this was achieved in Chapter 17 of this book.
⁵ Victoria, Government, *Protecting the Environment.*
⁶ Commonwealth of Australia, *Decade of Landcare Plan.* See also John Bradsen in Chapter 15 of this book.

17   Changing Cultures in the Farming Community

¹ S. N. Stone, 'Rural Communities: Facility Development, Attitudes and Self-Determination in the Gordon Shire'.
² For example see Don Paarlberg, *American Farm Policy.*
³ See Peter Small's plea in Chapter 14 of this book.
⁴ See Evan Walker's account of this commitment from government in Chapter 16 of this book.

18   Conclusion: Environmental Values, Knowledge and Action

¹ J. Baird Callicott and Roger T. Ames, 'The Asian tradition as a conceptual resource for environmental philosophy', p. 4.
² A. W. Watts, *Nature, Man and Woman*, p. 33.
³ Callicott and Ames, 'The Asian tradition', p. 6.
⁴ See F. Capra and D. Steindl-Rast, *Belonging to the Universe.*
⁵ Lee Talbot, 'The world's conservation strategy', pp. 266–7.
⁶ World Commission on Enviroment and Development, *Our Common Future*, p. 43.
⁷ Patricia H. Hynes, *The Recurring Silent Spring.*
⁸ J. Innes, 'Knowledge and action: Making the link', pp. 86–92.

[9] Innes, 'Knowledge and action', p. 87.
[10] *Environmental Research in Australia: A Review*, p. 2, quoted in Johnston and Berman, Chapter 8.
[11] D. Pirages and P. Ehrlich, *Ark II: Social Response to Environmental Imperatives*.
[12] R. E. Dunlap and K. D. Van Liere, 'Commitment to the dominant social paradigm and concern for environmental quality', pp. 1013–28.
[13] See M. P. Maloney and P. Ward, 'Ecology: Let's hear from the people', pp. 583–6.
[14] D. B. Gray, *Ecological Beliefs and Behaviour: Assessment and Change*.
[15] M. Fischbein and I. Ajzen, quoted in Gray, *Ecological Beliefs and Behaviour: Assessment and Change*, p. 131.
[16] M. Fischbein and I. Ajzen, *Belief, Attitude, Intention and Behaviour*.
[17] For a discussion of theories of attitude change see: Shelley Chaiken and Charles Stangor, 'Attitudes and attitude change', pp. 575–630; R. E. Petty and J. T. Cacioppo, *Attitudes and Persuasion: Classical and Contemporary Approaches*; and J. Cooper and R. T. Croyle, 'Attitudes and attitude change', pp. 395–426.
[18] J. Innes, *Knowledge and Public Policy*.
[19] J. Habermas, *The Theory of Communicative Action, Volume 1: Reason and the Rationalisation of society*. See also R. Bernstein, *Habermas and Modernity*.
[20] See the discussion of themes in Aboriginal poetry in S. Falkiner, *The Writer's Landscape: Wilderness*. Titles such as 'A bushtucker story' and 'Seedcake dreaming' are common amongst Aboriginal paintings (see, for example, paintings by Old Mick Tjakamarra and J. W. Tjupurrula in the collection of the National Gallery of Victoria).
[21] M. Berman, *The Re-enchantment of the World*.
[22] Judith Wright, 'Biological man', p. 167.

Bibliography

Alexandra, J., 'Political, social, economic and environmental issues in Australian water management', in J. J. Pigram (ed.), *Water Allocation for the Environment,* University of New England Centre for Water Policy Research, Armidale, New South Wales, 1992, pp. 111–21.

Allwright, John, 'The environment and sustainable growth—the key role of farmers', *Australian Journal of Soil and Water Conservation,* vol. 5, no. 1, 1992, pp. 4–6.

Australian Academy of Science, Global Change: A Research Strategy for Australia, 1992–96, Canberra, 1992.

Australian Bureau of Statistics, *Australia's Environment: Issues and Facts,* ed. I. Castles, ABS Cat 4140.0, Canberra, 1992.

Australian Capital Territory, Parks and Conservation Service, *ACT Decade of Landcare Plan: Landcare a Project for All,* draft, Canberra, September 1991.

An Australian Prayer Book, Anglican Information Office, Sydney, 1978.

Australian Science and Technology Council (ASTEC), *Environmental Research in Australia: A Review,* AGPS, Canberra, 1990.

——, *Environmental Research in Australia: The Issues,* AGPS, Canberra, 1990.

——, *Environmental Research in Australia: Case Studies,* AGPS, Canberra, 1991.

——, *Setting Directions for Australian Research,* AGPS, Canberra 1990.

Australian Soil Conservation Council, *National Soil Conservation Strategy,* AGPS, Canberra, 1988.

Barr, N. F., and Cary, J. W., *Greening a Brown Land: The Australian Search for Sustainable Land Use,* Macmillan, Melbourne, 1992.

Benn, Stanley, 'Property', in Paul Edwards (ed.), *Encyclopaedia of Philosophy*, vol. 6, Macmillan, New York, 1967, pp. 491–4.

Berman, M., *The Re-enchantment of the World*, Bantam Books, New York, 1988 (first published Cornell University Press, 1981).

Bernstein, R. (ed.), *Habermas and Modernity*, Polity Press, Cambridge, 1985.

Biological Diversity Advisory Committee, *A National Strategy for the Conservation of Australia's Biological Diversity: Draft for Public Comment*, Department of Arts, Sport, the Environment and Territories, Canberra, 1992.

Blaikie, P., *The Political Economy of Soil Erosion in Developing Countries*, Longman, Harlow, Essex, 1985.

Blakeway, Denys, and Lloyd-Roberts, Sue, *Fields of Thunder*, Unwin Paperbacks, London, 1985.

Bleier, Ruth (ed.), *Feminist Approaches to Science*, Pergamon Press, New York, 1986.

Bloor, D., 'Durkheim and Mauss revisited: classification and the sociology of knowledge', in N. Stehr and V. Meja (eds), *Society and Knowledge*, Transaction Books, New Brunswick, New Jersey, 1984, pp. 51–75.

Blyth, M., and McCallum, A., 'Onsite costs of land degradation in agriculture and forestry', in A. Chisholm and R. Dumsday (eds), *Land Degradation: Problems and Policies*, Cambridge University Press, Cambridge, 1987, pp. 79–98.

Boardman, K., and Eckersley R., 'An Australian agenda', in I. Marsh (ed.), *The Environmental Challenge*, Longman Cheshire, Melbourne, 1991, pp. 25–57.

Bookchin, M., *The Modern Crisis*, New Society Publishers, Philadelphia, 1986.

——, *Towards an Ecological Society*, Black Rose Books, Montreal, 1980.

Bradsen, J. R., *Soil Conservation Legislation in Australia*, Report for the National Soil Conservation Program, AUP, 1988, pp. 83–99.

——, 'Perspectives on land degradation', *Environmental and Planning Law Journal*, vol. 8, no. 1, 1991, pp. 16–40.

——, 'Hard sticks or soft carrots', paper presented to the International Conference on Sustainable Land Management, Hawkes Bay District Council, Napier, New Zealand, 1991.

——, and Fowler, R., 'Land degradation: legal issues and institutional constraints', in A. Chisholm and R. Dumsday (eds), *Land Degradation: Problems and Policies*, Cambridge University Press, Cambridge, 1987, pp. 129–67.

Brady, Veronica, 'Called by the land to enter the land', in Catherine

Hammond, *Creation Spirituality and the Dreamtime*, Millennium Books, Newtown, 1991, pp. 35–49.

Briassoulis, Helen, 'Theoretical orientations in environmental planning', *Environmental Management*, vol. 13, no. 4, pp. 381–92.

Bring, M., and Wayembergh, J., *Japanese Gardens: Design and Meaning*, McGraw-Hill, New York, 1981.

Brueggeman, Walter, *The Land: Overtures to Biblical Theology*, Fortress Press, Philadelphia, 1989.

Burch, G., Graetz, D., and Noble, I., 'Biological and physical phenomena in land degradation', in A. Chisholm and R. Dumsday (eds), *Land Degradation: Problems and Policies*, Cambridge University Press, Cambridge, 1987, pp. 27–48.

Buringh, P., and van Heemst, H. D., 'Potential world food production', in H. Linnemann, J. de Hoogh, M. A. Keyzer, and H. D. van Heemst (eds), *MOIRA: Model of International Relations in Agriculture* (Contr. Econ. Anal. no. 124), Elsevier North-Holland, Amsterdam, 1979, pp. 19–72.

Callicott, J. Baird, and Ames, Roger T., 'Introduction: The Asian traditions as a conceptual resource for environmental philosophy', in J. Baird Callicott and Roger T. Ames (eds), *Nature in Asian Traditions of Thought: Essays in Environmental Philosophy*, State University of New York Press, Albany, New York, 1989, pp. 1–21.

Capra, Fritjof, *The Tao of Physics*, Flamingo, London, 1983.

——, *The Turning Point: Science Society and the Rising Culture*, Flamingo, London, 1987.

——, and Steindl-Rast, D., *Belonging to the Universe*, Harper, San Francisco, 1991.

Carson, R. L., *Silent Spring*, Fawcett, Sydney, 1981 (first published by Houghton Mifflin, 1962).

Carter, V. G., and Dale, T. D., *Topsoil and Civilization* (revised edition), University of Oklahoma Press, Norman, Oklahoma, 1974.

Centre for Policy Research, University of Wollongong, *University Research Infrastructure: An Indicative Study, A study commissioned by the Australian Vice-Chancellors' Committee*, Wollongong, 1992.

Chaiken, Shelley, and Stangor, Charles, 'Attitudes and attitude change', *Annual Review of Psychology*, vol. 38, 1987, pp. 575–630.

Chartres, C., 'Australia's land resources at risk', in A. Chisholm and R. Dumsday (eds), *Land Degradation: Problems and Policies*, Cambridge University Press, Cambridge, 1987, pp. 7–26.

Chiew, F. H. S., and McMahon, T. A., 'Improved modelling of the groundwater processes in HYDROLOG', *Proceedings of the International Hydrology and Water Resources Symposium, Perth*, I.E. Aust. Nat.

Conf. Publ. 91/22(2), Institution of Engineers Australia, Canberra, 1991, pp. 492–7.

——, and O'Neill, I. C., 'Estimating groundwater recharge using an integrated surface and groundwater modelling approach', *Journal of Hydrology*, vol. 131, 1992, pp. 151–86.

Chisholm, A., 'Abatement of land degradation; regulations vs. economic incentives', in *Land Degradation: Problems and Policies*, A. Chisholm and R. Dumsday (eds), Cambridge University Press, Cambridge, 1987, pp. 223–47.

Choi, C. J., 'The P'ungsui theory (geomancy) and a Korean view of land', *Land Studies*, vol. 1, no. 4, 1990, pp. 30–41.

Christie, E. K., 'Foreword', in E. K. Christie (ed.), *Desertification of Arid and Semiarid Natural Grazing Lands*, Griffith University School of Australian Environmental Studies, Brisbane, 1981, pp. 1–5.

Clarke, H. R., Chisholm, A. H., Edwards, G. W., and Kennedy, J. O. S., *Immigration, Population Growth and the Environment*, AGPS, Canberra, 1990.

Commonwealth Department of Employment, Education and Training, *Australian Research Grants Scheme, Marine Sciences and Technologies Grants Scheme, National Research Fellowships Scheme, Queen Elizabeth II Fellowships Scheme: Grants and Fellowships Awarded 1988*, AGPS, Canberra, 1989.

Commonwealth Department of Environment, Housing and Community Development, *Commonwealth and State Government Collaborative Soil Conservation Study 1975–77 Report 1: A Basis for Soil Conservation Policy in Australia*, AGPS, Canberra, 1978.

Commonwealth Department of Primary Industries and Energy, *Third National Conservation Report*, AGPS, Canberra, 1989.

Commonwealth Department of Prime Minister and Cabinet, *Ecologically Sustainable Development (ESD), A Commonwealth Discussion Paper*, AGPS, Canberra, 1990.

Commonwealth of Australia, *Decade of Landcare Plan: Commonwealth Component*, AGPS, Canberra, 1991.

Condon, R. W., 'Pastoralism', in J. Messer and G. Mosely (eds), *What Future for Australia's Arid Lands?*, Australian Conservation Foundation, Hawthorn, Victoria, 1983, pp. 54–60.

Conservation Commission of the Northern Territory, *Northern Territory Decade of Landcare Plan*, Palmerston, Northern Territory, 1992.

Cooper, J., and Croyle, R. T., 'Attitudes and attitude change', *Annual Review of Psychology*, vol. 35, 1984, pp. 395–426.

Crane, W., 'The role of trees in sustainable agriculture', National Conference, Albury, New South Wales, September, 1991.

Crosby, Alfred W., *Ecological Imperialism: The Biological Expansion of Europe, 900–1900,* Cambridge University Press, Cambridge, 1986.

Daly, Herman E., and Cobb, John B. Jr, *For the Common Good,* Beacon Press, Boston, 1989.

Davidson, B., 'Comments', in A. Chisholm and R. Dumsday (eds), *Land Degradation: Problems and Policies,* Cambridge University Press, Cambridge, 1987, pp. 357–62.

Devall, Bill, and Sessions, George, *Deep Ecology: Living as if Nature Mattered,* Peregrine Smith, Salt Lake City, 1985.

Diamond, J., and Orenstein, G. (eds), *Reweaving the World,* Sierra Club Books, San Francisco, 1990.

Donald, C. M., and Prescott, J. A., 'Trace elements in Australian crop and pasture production, 1924–1974', in D. J. D. Nicholas and A. R. Egan (eds), *Trace Elements in Soil–Plant–Animal Systems,* Academic Press, New York, 1975, p. 7.

Dorn, H., *The Geography of Science,* Johns Hopkins University Press, Baltimore, 1991.

Dumsday, R., 'Contributions from the social sciences', in A. Chisholm and R. Dumsday (eds), *Land Degradation: Problems and Policies,* Cambridge University Press, Cambridge, 1987, pp. 315–34.

Dunlap, R. E., and Van Liere, K. D., 'Commitment to the dominant social paradigm and concern for environmental quality', *Social Science Quarterly,* vol. 65, 1978, pp. 1013–28.

Eberhardt, J., McDonald, D., and Papadopoulos, C., *Economic and Financial Issues: Water 2000 Consultant's Report 3,* AGPS, Canberra, 1983.

Ecologically Sustainable Development Working Group Chairs, *Intersectoral Issues Report,* AGPS, Canberra, 1992.

Economic Planning Advisory Council, *Managing Australia's Natural Resources,* Council Paper no. 49, AGPS, Canberra, 1992.

Egan, A. R., 'Animal production—leading the recovery', *Animal Production in Australia,* vol. 19, 1992, pp. xvii–xxi.

Ehrenfeld, David, *The Arrogance of Humanism,* Oxford University Press, New York, 1981.

Ehrlich, P., *The Population Bomb,* Ballantine, New York, 1962.

Elliot, Robert, 'Future generations, Locke's proviso and libertarian justice', *Journal of Applied Philosophy,* vol. 3, 1986, pp. 217–27.

Emel, J., and Peet, R. 'Resource management and natural hazards', in R. Peet and N. Thrift (eds), *New Models in Geography 1,* Unwin Hyman, London, 1989, pp. 49–76.

Falkiner, S., *The Writer's Landscape: Wilderness,* Simon and Schuster, Sydney 1992.

Faludi, Andreas, 'Three paradigms of planning theory', Paper for the Planning Theory Conference, held at the Oxford Polytechnic, Department of Town Planning, April 1981.

Field, Barron, *Geographical Memoirs of New South Wales (by various hands)*, John Murray, London, 1825.

Finlayson, B. L., and Bird, J. F., *Initial Investigation into the Extent and Nature of the Current Sedimentation Problem on the Lower Snowy River*, Department of Water Resources, Melbourne, 1989.

Fischbein, M., and Ajzen, I., *Belief, Attitude, Intention and Behaviour*, Addison-Wesley, Reading, Massachusetts, 1975.

Fisher, L. K., 'Discussion of trade-offs among soil conservation, energy use, experts and environmental quality', in H. G. Halcrow, E. O. Heady; and M. L. Cotner (eds), *Soil Conservation Policies, Institutions and Incentives*, Soil Conservation Society of America, Ankery, Iowa, 1982, pp. 274–9.

Forman, P., 'Kausalitat, Anschaulichkeit and Individualitat, or how cultural values prescribed the character and lessons ascribed to quantum mechanics', in N. Stehr and V. Meja (eds), *Society and Knowledge*, Transaction Books, New Brunswick, New Jersey, 1984, pp. 333–47.

Fox, Mathew, *Original Blessing: A Primer in Creation Spirituality*, Beare & Company, Santa Fe, New Mexico, 1983.

Fox, Warwick, *Towards a Transpersonal Ecology*, Shambhala, Boston, 1990.

French, R. J., and Schultz, J. E., 'Water use efficiency of wheat in a Mediterranean-type environment: II Some limitations of efficiency', *Australian Journal of Agricultural Research*, vol. 35, 1984, pp. 765–75.

Galloway, R. W., 'Natural and anthropogenic erosion and their relative impact in the arid zone', in J. Messer and G. Mosely (eds), *What Future for Australia's Arid lands?*, Australian Conservation Foundation, Hawthorn, Victoria, 1983, pp. 45–6.

Garnett, Stephen (ed.), *Threatened and Extinct Birds of Australia*, Royal Australasian Ornithologists Union, Moonee Ponds, Victoria, 1992.

Gelburd, D. E., 'Managing salinity: Lessons from the past', *Journal of Soil and Water Conservation*, vol. 40, no. 4, 1985, pp. 329–31.

Gray, D. B., *Ecological Beliefs and Behaviour: Assessment and Change*, Greenwood Press, Westport, Connecticut, 1985.

Gwaganad of Haada Gwaii (the Queen Charlotte Islands), 'Speaking for the earth: the Haida way', in Judith Plant (ed.), *Healing the Wounds: The Promise of Ecofeminism*, New Society Publishers, Philadelphia, 1989, pp. 76–9.

Habermas, J., *The Theory of Communicative Action, Volume 1: Reason and the Rationalisation of Society* (trans. T. McCarthy), Beacon Press, Boston, 1984.

Hannam, R. J., and Reuter, D. J., 'Review: Trace element nutrition of pasture', in J. H. Wheeler, C. J. Pearson and G. E. Robards (eds), *Temperate Pastures: Their Production, Use and Management*, CSIRO, Melbourne, 1987, p. 175.

Haraway, Donna, 'Animal sociology and a natural economy of the body politic', *Signs*, vol. 4, 1978, pp. 21–60.

Harding, Sandra, *The Science Question in Feminism*, Cornell University Press, Ithaca, New York, 1986.

Hardison, O. B., *Disappearing through the Skylight: Culture and Technology in the Twentieth Century*, Viking, New York, 1989.

Hart, B. T., 'Water quality management principles', in B. T. Hart (ed.), *Water Quality Management: Monitoring Programs and Diffuse Runoff*, Chisholm Institute of Technology Water Studies Centre, Melbourne, 1982, pp. 1–5.

Hartley, P. R., and Porter, M. G., 'A green thumb for the invisible hand', in I. Marsh (ed.), *The Environmental Challenge*, Longman Cheshire, Melbourne, 1991, pp. 243–68.

Hohfeld, W. N., *Fundamental Legal Conceptions as Applied in Legal Reasoning*, Yale University Press, New Haven, 1919.

Honoré, A. M., 'Ownership', in A. G. Guest (ed.), *Oxford Essays in Jurisprudence* (first series), Oxford University Press, Oxford, 1961, pp. 107–47.

Hooker, C. A., 'Responsibility, ethics and nature', in D. E. Cooper and J. A. Palmer, (eds), *The Environment in Question*, Routledge, London, 1992, pp. 147–64.

Hurley, F. T., Fitzgerald, B. C., Harvey, J. T., and Oppenheim, P. P., *Cropping and Conservation, A Survey of Cultivation Practices in Victorian Grain-Growing Areas*, Ballarat College of Advanced Education, Ballarat, Victoria, 1985.

Hynes, P. H., *The Recurring Silent Spring*, Pergamon Press, New York, 1989.

Illich, I., *De-Schooling Society*, Calder and Boyars, London, 1971.

Innes, J., 'Knowledge and action: Making the link', *Journal of Planning, Education and Research*, 1987, pp. 86–92.

——, *Knowledge and Public Policy*, Transaction Publishers, New Brunswick, New Jersey, 1990.

Irvine, J., 'Australian government funding of academic and related research—the international comparison', *Search*, vol. 21, 1990, pp. 88–95.

Jackson, F., 'How not to fund research', *Eureka Street*, vol. 3, no. 1, pp. 16–18.

Jameson, B., *Movement at the Station*, Collins, Sydney, 1987.

Jenkin, J. J., 'Salinity problems in Australia', in B. T. Hart (ed.), *Water Quality Management: Monitoring Programs and Diffuse Runoff*, Chisholm Institute of Technology Water Studies Centre, Melbourne, 1982, pp. 141–52.

Keary, J., 'Victoria's new approach to water development', in J. J. Pigram (ed.), *Water Allocation for the Environment*, University of New England Centre for Water Policy Research, Armidale, New South Wales, 1992, pp. 135–41.

Keller, Evelyn Fox, 'Contending with a masculine bias in the ideals and values of science', *Chronicle of Higher Education*, vol. 2, October 1985, p. 96.

——, *Reflections on Gender and Science*, Yale University Press, New Haven, Connecticut, 1985.

Kim, Kwi-Gon, *Report of Working Group Two on Urbanization and Environmental Change, the Workshop on Environmentally Sound and Sustainable Development*, Sponsored by the United Nations Development Programme, the Economic and Social Commission for Asia and the Pacific, and the Ministry of Environment of Korea, 27–29 June 1991.

Kirkby, M. J., 'The problem', in M. J. Kirkby and R. P. C. Morgan (eds), *Soil Erosion*, Wiley, Chichester, 1980, pp. 1–16.

Koh, J. K., 'Problems in the New Town Development Plan in Korea and a call for human ecological approach', *Environment and Landscape*, vol. 38, 1990, pp. 62–8.

Knudtson, P., and Suzuki, D., *Wisdom of the Elders*, Allen and Unwin, Sydney, 1992.

Korea, Ministry of Environment, *The National Declaration on Environmental Conservation*, 5 June 1992.

Lakoff, S. A., 'Scientists, technologists and political power', in I. Spiegel-Rosing and D. de S. Price (eds), *Science, Technology and Society: A Cross-Disciplinary Perspective*, Sage, London, 1977, pp. 355–91.

Land Conservation Council, *Rivers and Streams: Special Investigation*, Melbourne, 1983.

Larkins, F. P., 'Science and technology policies in Australia: Planning for the future', Address to the New Zealand Institute of Chemistry 1991 Jubilee Conference, University of Canterbury, Christchurch, 1991.

Layton, E., 'Conditions of technological development', in I. Spiegel-Rosing and D. de S. Price (eds), *Science, Technology and Society: A Cross-Disciplinary Perspective*, Sage, London, 1977, pp. 197–222.

Lee, Sang-Don, Kim, Kwi-Gon, and Lee, Sang-Gon, *National Report of the Republic of Korea to UNCED*, April 1992.

Leopold, Aldo, *A Sand County Almanac*, Oxford University Press, New York, 1949.

Lilburne, Geoffrey R., *A Sense of Place: A Christian Theology of the Land*, Abingdon Press, Nashville, 1989.

Lim, Ja-Hae, 'Tradition in Korean society: continuity and change', *Korea Journal*, vol. 31, no. 3, 1991, pp. 00.

Lloyd, Genevieve, *The Man of Reason*, Methuen, London, 1984.

Locke, John, *Two Treaties of Government*, ed. Peter Laslett, Mentor Books, New York, 1965.

Loomis, R. S., and Connor, D. J., *Crop Ecology: Productivity and Management in Agricultural Systems*, Cambridge University Press, Cambridge, 1992.

Lovejoy, S. B., and Napier, T. L., 'Conserving soil: Sociological insights', *Journal of Soil and Water Conservation*, vol. 41, 1986, pp. 304–10.

Mabbutt, J. A., *Desertification in Australia*, Water Research Foundation, Kingsford, New South Wales, 1978.

MacGuire, Daniel C., 'Values', in A. Richardson and J. Bowden (eds), *A New Dictionary of Christian Theology*, SCM Press, London, 1983.

MacLeod, R., 'Changing perspectives in the social history of science', in I. Spiegel-Rosing and D. de S. Price (eds) *Science, Technology and Society: A Cross-Disciplinary Perspective*, Sage, London, 1977, pp. 148–95.

Macpherson, C. B., *The Political Theory of Possessive Indivdualism: Hobbes to Locke*, Oxford University Press, Oxford, 1962.

Maloney, M. P., and Ward, P., 'Ecology: Let's hear from the people', *American Psychologist*, vol. 28, 1973, pp. 583–6.

Manes, Christopher, *Green Rage: Radical Environmentalism and the Unmaking of Civilisation*, Little, Brown and Company, Boston, 1990.

Mann, D. E., 'Environmental learning in a decentralized political world', *Journal of International Affairs*, vol, 44, no. 2, 1991, pp. 330–1.

Martin, B., *Scientific Knowledge in Controversy*, State University of New York Press, Albany, New York, 1991.

Massy, Charles, *The Australian Merino*, Viking, O'Neil, Melbourne, 1990.

Mathews, Freya, *The Ecological Self*, Routledge, London, 1991.

McDonagh, Sean, *To Care for the Earth: A Call to a New Theology*, Beare & Co., Sante Fe, 1986.

McKibben, B., *The End of Nature*, Penguin, London, 1990.

McMahon, T. A., and Finlayson, B. L., 'Australian surface and groundwater hydrology: regional characteristics and implications', in J. J. Pigram and B. P. Hooper (eds), *Water Allocation for the Environment*, Centre for Water Policy Research, University of New England, Armidale, New South Wales, 1992, pp. 21–40.

——, Haines, A. T., and Srikanthan, R., *Global Runoff: Continental Comparisons of Annual Flows and Peak Discharges*, Catena Verlag, Cremlingen-Destedt, 1992.

McPhail, I. R., and Young, E. M., 'Water for the environment in the Murray–Darling Basin', in J. J. Pigram (ed.), *Water Allocation for the Environment*, University of New England Centre for Water Policy Research, Armidale, New South Wales, 1992, pp. 191–210.

McPhee, J., *The Control of Nature*, Pimlico, London, 1991.

Merchant, Carolyn, *The Death of Nature*, Harper and Row, New York, 1980.

——, *Radical Ecology*, Routledge, London, 1992.

Michiko, Ishimure, *Paradise in the Sea of Sorrow: Our Minamata Disease* (originally published 1972 as *Kugai Jodo: Waga Minamato-byo*), trans. Livia Monnet, Yamayuchi Publishing House, Tokyo, 1990.

Miller, Doug, 'What the polls tell us', *Women and Environments*, Winter/Spring 1991, pp. 68–9.

Mitchell, J. K., and Bubenzer, G. D., 'Soil loss estimation', in M. J. Kirkby and R. P. C. Morgan (eds), *Soil Erosion*, Wiley, Chichester, 1980, pp. 17–62.

Moltmann, Jurgen, *God in Creation: An Ecological Doctrine of Creation*, Gifford Lectures 1984–5, SCM Press, London, 1985.

Morgan, R. P. C., 'Implications', in M. J. Kirkby and R. P. C. Morgan (eds), *Soil Erosion*, Wiley, Chichester, 1980, pp. 253–301.

Mortimore, M., *Adapting to Drought*, Cambridge University Press, Cambridge, 1989.

Mulkay, M. J., 'Sociology of the scientific research community', in I. Spiegel-Rosing and D. de S. Price (eds) *Science, Technology and Society: A Cross-Disciplinary Perspective*, Sage, London, 1977, pp. 93–148.

Murray–Darling Basin Commission, *Managing Australia's Heartland*, Canberra, 1993.

Murray–Darling Basin Ministerial Council, *Murray–Darling Basin Natural Resources Management Strategy*, Canberra, 1990.

Naess, A., 'Identification as a source of deep ecological attitudes', in M. Tobias (ed.), *Deep Ecology*, Avant Books, San Diego, 1985.

—— 'The shallow and the deep, long-range ecology movement', *Inquiry*, vol. 16, 1973, pp. 95–100.

——, *Ecology, Community and Lifestyle* (translated and revised by David Rothenberg), Cambridge University Press, Cambridge, 1989.

——, and Sessions, G., 'Basic principles of deep ecology', *Ecophilosophy*, 1984, vol. 6, pp. 3–7.

Nash, Roderick, *The Rights of Nature: A History of Environmental Ethics*, Primavera Press, Sydney, 1990.

Neville, S., *The Australian Environment: Taking Stock and Looking Ahead*, Australian Conservation Foundation, Fitzroy, Victoria, 1990.

Norton, Bryan G., *The Preservation of Species: The Value of Biological Diversity*, Princeton University Press, Englewood Cliffs, New Jersey, 1988.

Nozick, R., *Anarchy, State, and Utopia*, Basic Books, Englewood Cliffs, New Jersey, 1974.

Ongley, E. D., 'Information requirements for water quality management: a reflective appraisal of present practices and future requirements', in B. T. Hart (ed.), *Water Quality Management: Monitoring Programs and Diffuse Runoff*, Chisholm Institute of Technology Water Studies Centre, Melbourne, 1982, pp. 7–21.

O'Riordan, T., 'The challenge for environmentalism', in R. Peet and N. Thrift (eds), *New Models in Geography 1*, Unwin Hyman, London, 1989, pp. 77–102.

Organisation for Economic Cooperation and Development, *Reviews of National Science and Technology Policy: Australia*, OECD, Paris, 1986.

——, Directorate for Science Technology and Industry, *Environmental Change and Science and Technology Institutions, The Experience of Selected OECD Countries*, Synthesis report DSTI/STP (91) 22, Paris, 1991.

Ortoland, L., *Environmental Planning and Decision Making*, John Wiley, New York, 1984.

Paarlberg, Don, *American Farm Policy*, John Wiley, New York, 1986.

Parliament of Victoria Salinity Committee, *Salt of the Earth: Final Report on the Causes, Effects and Control of Land and River Salinity in Victoria*, Government Printer, Melbourne, 1984.

Passmore, John, *Man's Responsibility for Nature: Ecological Problems and Western Traditions*, Duckworth, London, 1974.

Petty, R. E., and Cacioppo, J. T., *Attitudes and Persuasion: Contemporary and Classical Approaches*, William C. Brown, Dubuque, Iowa, 1981.

Pirages, D., and Ehrlich, P., *Ark II: Social Response to Environmental Imperatives*, Viking, New York, 1974.

Plant, Judith (ed.), *Healing the Wounds: The Promise of Ecofeminism*, New Society Publishers, Philadelphia, 1989.

Porter, J. W., and McMahon, T. A., *The Monash Model: User Manual for Daily Program HYDROLOG*, Civil Engineering Research Report, 2/1976, Monash University, Melbourne, 1976.

Price, D. de S., 'Is technology historically independent of science?', *Technology and Culture*, vol.6, 1965, pp. 553–67.

Rasmussen, W. D., 'History of soil conservation institutions and incentives', in H. G. Halcrow, E. O. Heady, and M. L. Cotner (eds), *Soil Conservation Policies Institutions and Incentives*, Soil Conservation Society of America, Ankery, Iowa, 1982, pp. 3–18.

Ravetz, J. R., 'Criticisms of science', in I. Spiegel-Rosing and D. de S. Price (eds), *Science, Technology and Society: A Cross-Disciplinary Perspective*, Sage, London, 1977, pp. 71–89.

Reader, J., *Man on Earth*, Collins, London, 1988.

Reeve, Andrew, *Property*, Macmillan, London, 1986.

Reeve, I. J., Patterson, R. A., and Lees J. W., *Land Resources: Training Toward 2000*, University of New England Rural Development Centre, Armidale, New South Wales, 1989.

Resource Assessment Commission, *Kakadu Conservation Zone Inquiry*, AGPS, Canberra, 1991.

——, *Forest and Timber Inquiry*, AGPS, Canberra, 1992.

Rickson, R., Saffigna, P., Vanclay, F., and McTainsh, G., 'Social bases of farmers' responses to land degradation', in A. Chisholm and R. Dumsday (eds), *Land Degradation: Problems and Policies*, Cambridge University Press, Cambridge, 1987, pp. 187–200.

Roberts, B. R., 'Land ethics—A necessary addition to Australian values', Soil Degradation Conference, Australian National University, Canberra, 1984.

Roberts, Brian, *Land Care Manual*, New South Wales University Press, Kensington, New South Wales, 1992.

Robertson, G., 'Contributions from the physical and biological sciences', in A. Chisholm and R. Dumsday (eds), *Land Degradation: Problems and Policies*, Cambridge University Press, Cambridge, 1987, pp. 305–14.

Rothbard, M., *For a New Liberty*, New York, Macmillan, 1972.

Rural Water Commission of Victoria, *Inquiry into Water Allocations in Northern Victoria*, Melbourne, 1984.

Sagoff, Mark, *The Economy of the Earth*, Cambridge University Press, Cambridge, 1988.

Salinity Bureau, *Victorian Salinity Program: First Annual Review, 1987–1988*, Department of Premier and Cabinet, Melbourne, 1988.

——, *Victorian Salinity Program: Second Annual Review, 1988–1989*, Department of Premier and Cabinet, Melbourne, 1989.

Schaefer, M., 'The federal research puzzle: Making the pieces fit', *Environment*, vol. 33, no. 9, 1991, p. 18.

Seagram, G. N., and Lendron, R. J., *Furnishing the Mind: Comparative Study of Cognitive Development in Central Australian Aborigines*, Harcourt Brace Jovanovich, New York, 1980.

Simonds, J. O., *Landscape Architecture: The Shaping of Man's Natural Environment*, McGraw-Hill, New York, 1961.

Singer, P., 'Environmental values', in I. Marsh (ed.), *The Environmental Challenge*, Longman Cheshire, Melbourne, 1991, pp. 3–24.

Smith, D. G., *Continent in Crisis*, Penguin, Ringwood, Victoria, 1990.

Smith, D. I., and Finlayson, B. L., 'Water in Australia: its role in environmental degradation', in R. L. Heathcote and J. A. Mabbutt (eds), *Land, Water and People: Geographical Essays in Australian Resource Management*, Allen and Unwin, Sydney, 1988, pp. 7–48.

South Australia, Department of Agriculture, Soil and Water Conservation Branch, *Decade of Landcare Plan for South Australia: Towards Sustainable Land Resource Management*, Adelaide, 1991.

Spiegel-Rosing, I., 'The study of science, technology and society (SSTS): recent trends and future challenges', in I. Spiegel-Rosing and D. de S. Price (eds), *Science, Technology and Society: A Cross-Disciplinary Perspective*, Sage, London, 1977, pp. 7–42.

Standing Committee on Agriculture, *Sustainable Agriculture*, Report of the Working Group on Sustainable Agriculture, SCA Technical Report Series no. 36, 1991.

Stanley, R. J., 'Soils and vegetation: an assessment of current status', in J. Messer and G. Mosely (eds), *What Future for Australia's Arid Lands?*, Australian Conservation Foundation, Hawthorn, Victoria, 1983, pp. 8–18.

Stone, Christopher, *Earth and Other Ethics: The Case for Moral Pluralism*, Harper & Row, New York, 1988.

Stone, S. N., Rural Communities: Facility Development, Attitudes and Self-Determination in the Gordon Shire, MA thesis, School of Social Sciences, LaTrobe University, 1977.

Talbot, Lee, 'The world's conservation strategy', *Environmental Conservation*, vol. 7, no. 4, 1980, pp. 266–7.

Tasmania, Department of Primary Industry and Fisheries, *The Tasmanian Decade of Landcare Plan*, Hobart, 1992.

Tillich, Paul, *Systematic Theology*, University of Chicago Press, Chicago, 3 volumes, 1951, 1957, 1963.

Tribe, Lawrence H., 'Ways not to think about plastic trees', in Lawrence H. Tribe, Corinne S. Schelling, and John Voss (eds), *When Values Conflict: Essays on Environmental Analysis*, Ballinger, Cambridge, Massachusetts, 1976, pp. 61–92.

Tudge, Colin, *Last Animals at the Zoo: How Mass Extinction Can Be Stopped*, Island Press, Washington, 1992.

United Kingdom Cabinet Office, Advisory Council on Science and Technology, *Environmental Research Programmes*, London, 1992.

University of Melbourne, *Research Report,* 1988, Melbourne, 1988.
——, *Research Report, 1989,* Melbourne, 1989.
Upstill, G., and Yapp, T., 'Offsite costs of land degradation', in A. Chisholm and R. Dumsday (eds), *Land Degradation: Problems and Policies,* Cambridge University Press, Cambridge, 1987, pp. 99–109.
Uren, N. C., 'The management of soil organic matter for sustainable agriculture', *Agricultural Science,* vol. 4, 1991, pp. 45–8.
Victoria, Department of Conservation and Environment, *The Victorian Decade of Landcare Plan,* Melbourne, 1992.
Victoria, Government, *Protecting the Environment: A Conservation Strategy for Victoria,* Government Printer, Melbourne, 1987.
——, *Salt Action: Joint Action,* Government Printer, Melbourne, 1988.
Victorian Parliamentary Salinity Committee, *Salt of the Earth: Final Report on the Causes, Effects and Control of Land and River Salinity in Victoria,* Government Printer, Melbourne, 1984.
Waldron, Jeremy, 'What is private property?', *Oxford Journal of Legal Studies,* vol. 5, 1985, pp. 326–33.
Walker, A. B., 'Responses in the New Zealand meat and wool sector downturn', National Agricultural Outlook Conference, Canberra, 1992.
Wasson, R., 'Detection and measurement of land degradation processes', in A. Chisholm and R. Dumsday (eds), *Land Degradation: Problems and Policies,* Cambridge University Press, Cambridge, 1987, pp. 49–69.
Watson, W. D., Reynolds, R. G., Collins, D. J., and Hunter, R. D., *Agricultural Water Demand and Issues: Water 2000 Consultants Report 5,* AGPS, Canberra, 1983.
Watt, K. E. F., *Understanding the Environment,* Allyn and Bacon, Boston, 1982.
Watts, A. W., *Nature, Man and Woman,* Vintage Books, New York, 1970.
Western Australia, Department of Agriculture, and Soil and Land Conservation Council, *Decade of Landcare Plan, Western Australia,* South Perth, 1992.
White, Lynn Jnr, 'The historical roots of our ecological crisis', *Science,* vol. 155, 1967, pp. 1203–7.
Wigmore, Lionel, *Struggle for the Snowy,* Oxford University Press, Melbourne, 1968.
Williams, B. G., 'Salinity and waterlogging in the Murray–Darling Basin', in Australian Science and Technology Council, *Environmental Research in Australia: Case Studies,* AGPS, Canberra, 1991, pp. 87–120.
Williams, M. D., *Out of the Mist, Book II: The Story of Man's Mastery of his Physical Environment, and his Intellectual and Spiritual Development,* Oldham, Beddome and Meredith, Tasmania, 1948.

Williams, W. D., 'Water for sustainable resource management within a semi-arid continent', in J. J. Pigram (ed.), *Water Allocation for the Environment*, University of New England Centre for Water Policy Research, Armidale, New South Wales, 1992, pp. 11–17.

Willmot, E., *The Culture of Literacy*, Australian Institute of Aboriginal Studies, Canberra, 1982.

——, *The Boyer Lectures: Australia, The Last Experiment*, ABC Publications, Sydney, 1986.

——, *Education's Hidden Outcome: Tenth Oscar Mendelsohn Lecture*, Monash University, Melbourne, 1988.

Wilson, E. O., (ed), *Biodiversity*, National Academy Press, Washington, D.C., 1988.

Wittwer, S., 'New technology, agricultural productivity and conservation', in H. G. Halcrow, E. O. Heady and M. L. Cotner (eds), *Soil Conservation Policies Institutions and Incentives*, Soil Conservation Society of America, Ankery, Iowa, 1982, pp. 201–15.

Woodruff, N. P., and Siddoway, F. H., 'A wind erosion equation', *Proceedings, Soil Science Society of America*, vol. 29, 1965, pp. 602–8.

Woods, L. E., *Land Degradation in Australia*, AGPS, Canberra, 1983.

World Commission on Environment and Development, *Our Common Future*, Oxford University Press, Oxford, 1987.

World Council of Churches, Commission on Church and Society, *Faith, Science and the Future*, 2 vols, Geneva, 1980.

Wright, Judith, 'Biological man', in G. Seddon and M. Davis (eds), *Man and Landscape in Australia: Towards an Ecological Vision*, AGPS, Canberra, 1976.

——, 'Eroded hills', in *Collected Poems*, Angus and Robertson, Melbourne, 1971, p. 83.

Yencken, David, 'The links between issue identification and action: Dilemmas for scientists', in J. N. Coles and J. M. Drew (eds), *Australia and the Global Environmental Crisis: Looking for Peaceful Solutions*, Academy Press, Canberra, 1992, pp. 253–71.

Young, J., *Post Environmentalism*, Belhaven Press, London, 1990.

Zimmerman, M. E., (ed), *Environmental Philosophy: From Animal Rights to Radical Ecology*, Prentice Hall, Englewood Cliffs, NJ.

Index

Aboriginal, 236; land rights, 10, 17, 58; land tenure, 13; people, 23–5, 42, 46

Aborigines, 9, 13, 15, 18, 43

action, 60–2, 79, 80, 82, 217; co-operative, 209–16; discretionary, 181

agriculture, 113, 117, 121, 136, 139–40, 142, 144, 146–7, 149–50, 163, 166–7, 176, 195, 204; agricultural business, 143; agricultural commodities, 127, 142, 152; organic methods, 141; productivity, 136, 138–9, 141–2, 144–5, 147, 154, 206; surplus production, 142; sustainable, 4–5, 136, 138–9, 149–50, 171–2, 205–6, 222, 228; trade, 144; world grain production, 141–2

American Dustbowl, 178, 185, 224

anima mundi, see genius loci

anthropocentric, 20, 37–40, 45, 62–3, 72–3, 124, 138

Aral Sea, 113

attitudes, 60, 137, 177–9, 217, 223, 234; and behaviour, 3, 6, 177–9, 223, 234; change in, 2, 3, 177, 178, 224–5

Australia, 15–18, 40–1, 83–4, 86–7, 89–94, 96–7, 102–5, 120–1, 126–9, 135–9, 142–4, 167–71, 174–5, 182–90, 193–4, 202–7, 228–9; climate, 126, 152, 157–8; early settlement of, 13–14, 17, 37, 40, 139; landscapes, 41; soils, 4, 120, 126, 152; water, 120, 129

Australian Academy of Science (AAS), 92

Australian Bureau of Statistics (ABS), 86

Australian Conservation Foundation (ACF), 164, 183, 199–200

Australian Institute of Marine Science (AIMS), 88–90

Australian Research Council, 89, 97, 101–5, 229–30

Australian Science and Technology Council (ASTEC), 84, 86, 91, 94, 96, 98, 229

Bacon, Sir Francis, 14, 110

beliefs, 3, 15, 75, 136, 158, 221, 232–4; core, 1, 6, 222–3, 232–5; *see also* environmental values; values

biodiversity, 86, 172, 174–6, 178, 182, 184, 191–2, 222; loss of, 18

biological diversity, see biodiversity

biotechnology, 131

Birch, Charles, 31–2

Bophal, 113

Brundtland Commission, the, 221
Buddhism, 63, 65

Canada, 205
Carson, Rachel, 1
change, 149, 161–3, 173, 177–8, 223–5, 232–6; barriers to, 153, 159, 163; indicators of change, 145
China, 26, 63, 123
Christian/Christianity, 27, 30–3, 35–6, 38, 123, 178, 218–19
collapse of civilizations, 173–4, 195
'commons, the', 143, 180
Commonwealth Scientific and Industrial Research Organisation (CSIRO), 89, 96–7, 101–4
communication, 86, 209–15
communicative action, 234
community consultation, 197–8, 209, 211–14; *see also* consensual processes; dialogue; action, co-operative
Confucianism, 63, 65
consensual processes, 225, 233; *see also* community consultation; dialogue; action, co-operative
conservation, 17–19, 81, 122, 126–7, 130, 132, 166, 174–5, 179–80, 185–7, 191–2, 196, 205–6, 220, 230
conservation farming, *see* farming, conservation
conservation movement, *see* environmental movement
Conservation Strategy (Victoria), 199
Conventions on Climatic Change and Biodiversity, 70
Co-operative Research Centres (CRC), 105
cosmology, 10
creator/creation, 27–30, 34

deep ecology, 39–40, 42, 123
democracy, 180
development, 18, 21, 71, 173, 217, 220, 231
dialogue, 6, 28, 32, 223, 233; *see also*
community consultation; consensual processes; dialogue
Dominant Social Paradigm, 72, 232
dominium, 29–30, 33–4

Earth Summit, *see* UNCED
ecofeminism, 39–40, 42
Ecologically Sustainable Development, 84, 89–92
ecology, 31
economics, 123, 128–30, 144, 146, 172, 179–80, 182, 223–4; cost-benefit analysis, 179; discount rate, 122, 179; economic incentives, 225; externalities, 143–4, 179; markets, 179; pricing, 122; value, 179
Economic Planning Advisory Council (EPAC), Australia, 86
ecophilosophy, 39
ecosystem, 140, 159, 175–6, 179, 184
education, 149, 181, 196, 225; environmental, 80
Enlightenment, 14
environment, 5, 28, 47–8, 51, 53–4, 56, 57–9, 62, 66, 69, 71–3, 74, 79–81, 83, 111–12, 114, 126, 130–1, 146, 205; definition of, 79, 217–18, 220; scientific view of, 79
Environment Protection Agency (US), 224
environmental awareness, 66
environmental debates, 1
environmental groups, 107
environmental impact analysis, *see* environmental impact assessment
environmental impact assessment, 67, 89, 113–14, 143
environmental impacts, 144
environmental legislation, 67–9, 176
environmental management, 83–4, 91, 130, 212
environmental movement, 15, 70, 71, 115, 165
environmental problems, 2–4, 6, 9, 11, 67, 74, 79, 81, 93, 107, 112–14,

116, 121, 126, 146, 167–8, 195, 208–9, 211, 215, 217, 222–6, 228–31; climate change, 130–1, 147; ozone depletion, 130; pollution, 1, 3, 4, 18, 54, 67, 143, 145, 224; population, 25, 26, 29, 146–8, 150; species extinction, 175, 179; world population, 141–2, 147; deforestation, 113

environmental research, 5–6, 62, 79–86, 88, 91, 93–8, 102–3, 105–7, 113, 128, 223, 228–30; Australian, 86, 89; data, 85, 86–8, 93; funding, 85, 88, 93–9, 101–4; policy, 83–4, 86, 92, 101; priorities, 91, 98; *see also* research

Environmental Resources Information Network (ERIN), Australia, 87

environmental science, 68, 80–1, 117–18, 128, 130–2, 228, 230; *see also* science

environmental values, 9, 12, 49, 80, 90; *see also* beliefs; values

environmentalism/environmentalist, 18, 20, 62, 72, 108, 111, 114, 130, 166, 215

erosion, 121, 124, 125, 127–8, 209

ethic(s), 72, 130, 172, 177–8, 235; environmental, 1, 56, 136, 235; land ethic, 178, 215

farm(s), 202–5, 207, 212–14

farmers, 81, 116–17, 121–2, 126–7, 129, 137, 139, 151–3, 155–68, 170–2, 198, 202–6, 208–10, 213, 215–16, 223, 227, 233–5; multi-skilling of, 160; values, 202–5, 207, 212–13, 215

farming, 11, 44–6, 56, 120, 139, 141–3, 145, 148–51, 153, 155, 157, 160, 161, 165, 167, 202–5, 207; alternative conceptions of, 46, 149; communities, 207–8, 210; conservation farming, 129, 146, 154–8, 161, 163, 225, 231; European methods, 11, 139; ethos, 203–5, 207, 215–16; extension services for, 163, 181; irrigation, 129, 202–3, 208; laser

levelling, 129, 208; new technologies in, 208; organic methods, 141, 148–9, 206; problems in, 167, 168; stubble retention, 129; sustainable systems, 162, 164; tree planting, 129, 168–70, 184; whole farm planning, 129

First World, 26

food, 43, 141; European, 42, 45; indigenous, 11, 40, 42–5; production, 136, 138, 142, 144, 147–9, 166, 205

Friends of the Earth (FOE), 108

Genesis, Book of, 28–30, 32

genius loci, 41, 42, 235–6

geomancy, 68, 71; *see also* P'ung Sui

global warming, *see* environmental problems, climate change

Greek philosophy, 38

Green Revolution, 142

Greenpeace, 108

Greens, *see* environmental movement

Gulf War, 113

Habermas, Jürgen, 234

heritage, 9

Humanity: First, 10, 23–6; Second, 10, 23–6; Third, 10, 26

imago Dei, 28–9, 33–5

India, 112

Industrial Revolution, 13, 35

information, 162, 163, 226; transfer of, 163

investigation, 94, 100, 103

Japan, 20, 112

Judeo-Christian religion, 28–9, 218

justice, 27, 30, 32, 35

Kenya, 112

knowledge, 4, 60, 61, 73, 79, 81–3, 113, 117–18, 124, 135, 155, 166, 217, 223, 226–7, 230, 234; community-based, 3, 112, 227; expert, 3, 6, 111, 114, 223, 228; feminist,

229; scientific, 3, 5, 6, 116, 119, 124, 217, 230
Korea, 11, 59–75, 236

land, 37, 42–5, 47, 55–7, 63, 125, 127–8, 131–2, 139, 143, 170, 173–4, 179, 181, 183, 186, 191, 195, 198, 215–16, 235–6; as sacred, 9–10, 13, 15, 17, 25, 42; mother land, 41; perceptions of, 10–12, 17
Land Conservation Council (LCC), Victoria, 101
land degradation, 4, 5, 13, 81, 117, 119, 120–30, 132, 135, 140, 145, 151–3, 155, 166, 169, 171–4, 176, 179–81, 183, 187, 188, 190, 193, 207, 209, 221–2, 226, 229–30
land rehabilitation, 145, 230
Landcare, 96, 135, 137, 153, 160, 162, 164, 168, 172, 182–7, 193, 198–201, 225, 227, 231, 233–5; decade of, 183, 186, 199; effectiveness, 185–6; groups, 183, 185, 199
language, 12, 17–18, 32
law, see legislation
legislation, 172, 179–81, 187–92, 200, 224, 231; Conservation Reserve Program (US), 182; see also soil conservation legislation
Liberal-Democratic principles, 48, 52
libertarian, 54
liberty, 53–4, 56, 57
Love Canal, 112

Mabo High Court decision, 58
Malthus, Thomas, 141
Marx/Marxism, 61
Melbourne, 44
Menzies, Sir Robert, 16
Minamata Bay, 4, 112
modelling, 100–1
Moltmann, Jurgen, 29–30, 33, 35
Murray–Darling Basin, 86, 125, 168, 193, 197–8
Murray Darling Basin Commission, 89, 96–7, 104, 184, 197
Murray Darling system, see Murray

Darling Basin
Murray River, 195

National Environmental Policy Act (US), 224
National Farmers' Federation (NFF), Australia, 164, 183, 185
National Resource Information Centre (NRIC), Australia, 87
National Soil Conservation Program, 184
native food, see indigenous food
natural world, see nature
nature, 11, 14–15, 19–22, 23, 25–6, 29–35, 38–40, 44–6, 48, 59, 62, 68, 70–2, 74, 79, 110, 113, 116, 121, 123, 131–2, 173, 176, 179, 218–20, 232, 235–6; alienation from, 40, 44; re-enchantment of, 39
New Environmental Paradigm, 72, 232
New South Wales, 14, 121, 135, 200
New Zealand, 166, 174–5, 205
Noah's Ark, 11, 29, 41, 44

opinion polls, 2
Organisation for Economic Co-operation and Development, 89

paradigm, 60, 61; see also worldview
participation/participatory, 32, 35–6, 61
Permaculture, 46
Piaget, Jean, 24
planning, 60–1, 70, 72–3, 75, 80; community-led, 210–14; environmental, 80, 211
policy, government, 193–201
progress, see development
property, 11, 14, 47–55, 189–90; liberal theories of, 58; ownership, 49–54, 57–8; private, 14, 50, 126, 180–1; property rights, 11, 47–58, 130; regulation, 180–1; taxation, 51; theories of justification, 50, 52–3, 55–6
P'ung Sui, 63, 65–6, 71–4, 138, 236; see also geomancy

Queensland, 135

rain forests, 19
religion, 10, 15, 25–6, 28, 31, 36
research, 93–5, 98, 99, 103–4, 117–19, 124, 126–7, 149, 151–3, 158–9, 171, 178, 214, 227–9, 231; and development, 84–5, 88, 90–1, 118, 139, 162, 170; curiosity driven, 102, 104, 105, 106, 229; interdisciplinary, 80, 85, 90–1, 92, 95, 101–2, 104, 230; monitoring, 86–7, 94, 100, 103, 126; transfer of, 163, 223, 227–8; scientific, 99; see also environmental research
Resource Assessment Commission (RAC), 86
Rural Water Commission, Victoria, 129

salinity/salinization, 4, 86, 99–100, 121, 126, 129, 137, 145, 193–200, 202–3, 205, 210, 211, 214, 231
salinity programmes, 99, 137, 193, 199–202, 210, 214–15, 225, 227, 233–5
scholarship, 95, 100, 103
science, 1, 3, 6, 14, 31–2, 38–9, 73, 81, 93, 107–10, 112–19, 121, 123–4, 129–31, 172, 176–7, 217–18, 221, 223–6, 230–1; and technology, 16, 28, 31, 68, 79, 92, 231–2; critiques of, 108; feminist view of, 107–11; gendered basis of, 81, 109, 111, 115; Western, 68, 79, 109, 110, 113–14, 117, 127, 176, 231; see also environmental science
Seagram, Gavin, 24
Silent Spring, 1, 2
Smith, Adam, 14–16, 18, 20
Snowy Mountains Scheme, 16
soil, 124, 140, 145–6, 159, 182, 187, 188, 190
soil conservation legislation, 188–90; see also legislation
South Australia, 23, 137, 182, 184, 188, 192, 231
spirit of the land, see genius loci

spirit of place, see genius loci
spirituality, 42
stewardship, 29, 34–5
sustainability, 10, 17–18, 27, 30, 32, 35, 59, 62, 69, 70, 75, 136, 138, 143–4, 150–3, 159, 162, 164–6, 177, 180, 182, 186–9, 205–6, 209, 215, 217–18, 220–2; definitions of, 5, 172–3, 175, 220; measures of, 159
sustainable development, 19, 21, 60–1, 70, 74, 84–5, 173, 220–1
sustainable use, 174
Suzuki, David, 10, 23–4

Taoism, 63, 66, 73
Third World, 26
Tillich, Paul, 27
tree planting, see farming

UNCED Conference, 13, 20, 70
United States, 112, 128, 181–2, 185, 205
universities, 88–9, 99, 102–3, 118
urbanization, 73
utility/utilitarian, 13, 19, 46, 53, 55, 72

value free, 12, 28, 108
value systems, 1, 12, 71–2, 73
values, 3–6, 10–12, 27–8, 32, 36, 48, 50, 53, 57–8, 60, 75, 108, 135–7, 213, 217–18, 221–3, 232–3, 235; aesthetic, 56; change in, 35, 73, 232; communal, 30; core, 3, 222; Eastern, 11, 12; instrumental, 218; intrinsic, 38, 39, 40, 56, 218; religious, see religion; spiritual, 30; traditional, 10; traditional rural, see farmers, values; Western, 15, 19, 71, 220–1, 232
Victoria, 96, 99–101, 135, 137, 168–9, 193–200, 202–3, 205, 207, 209, 211, 213, 227

water, 129, 181, 191, 200, 215
Wealth of Nations, the, 14
Western Australia, 40, 135, 138, 187
wilderness, 45

women, 9, 34, 110–14, 210
World Conservation Strategy, 220
World Council of Churches, 30, 31
World Heritage, 90
worldview, 119, 219, 221; dis-
    enchanted, 38; dominant, 31;
dualistic, 39, 219–20; mechanistic,
    31, 219
Wright, Judith, 37, 236

Yin/Yang, 63, 71

Zen Buddhism, 63, 65

DATE DUE

MAY 0 4 2005

DEMCO, INC. 38-2931